Empowering the Public Sector with Generative AI

From Strategy and Design to Real-World Applications

Sanjeev Pulapaka
Srinath Godavarthi
Dr. Sherry Ding

Apress®

Empowering the Public Sector with Generative AI: From Strategy and Design to Real-World Applications

Sanjeev Pulapaka
Brambleton, VA, USA

Srinath Godavarthi
Stone Ridge, VA, USA

Dr. Sherry Ding
Herndon, VA, USA

ISBN-13 (pbk): 979-8-8688-0472-4
https://doi.org/10.1007/979-8-8688-0473-1

ISBN-13 (electronic): 979-8-8688-0473-1

Managing Director, Apress Media LLC: Welmoed Spahr
Acquisitions Editor: Celestin Suresh John
Development Editor: Laura Berendson
Editorial Project Manager: Gryffin Winkler

Cover designed by eStudioCalamar

Cover image designed by NicoElNino on iStock

Distributed to the book trade worldwide by Springer Science+Business Media New York, 1 New York Plaza, Suite 4600, New York, NY 10004-1562, USA. Phone 1-800-SPRINGER, fax (201) 348-4505, e-mail orders-ny@springer-sbm.com, or visit www.springeronline.com. Apress Media, LLC is a California LLC and the sole member (owner) is Springer Science + Business Media Finance Inc (SSBM Finance Inc). SSBM Finance Inc is a **Delaware** corporation.

For information on translations, please e-mail booktranslations@springernature.com; for reprint, paperback, or audio rights, please e-mail bookpermissions@springernature.com.

Apress titles may be purchased in bulk for academic, corporate, or promotional use. eBook versions and licenses are also available for most titles. For more information, reference our Print and eBook Bulk Sales web page at http://www.apress.com/bulk-sales.

Any source code or other supplementary material referenced by the author in this book is available to readers on GitHub. For more detailed information, please visit https://www.apress.com/gp/services/source-code.

If disposing of this product, please recycle the paper

Table of Contents

About the Authors

Sanjeev Pulapaka is Principal Solutions Architect at Amazon Web Services (AWS). He leads the development of AI/ML and Generative AI solutions for the US Federal Civilian team. Sanjeev has extensive experience in leading, architecting, and implementing high-impact technology solutions that address diverse business needs in multiple sectors (including commercial, federal, and state and local governments). He has published numerous blogs and white papers on AI/ML and is an active speaker and panelist at various industry conferences, including AWS Public Sector Summit and AWS re:Invent. Sanjeev has an undergraduate degree in engineering from the Indian Institute of Technology and an MBA degree from the University of Notre Dame.

Srinath Godavarthi has over 20 years of experience serving public sector customers and has held leadership positions with global technology and consulting companies, including Amazon and Accenture. In his previous roles, Srinath led cloud strategy, architecture, and digital transformation efforts for a number of federal, state, and local agencies. Srinath specializes in AI/ML technologies and has published over a dozen white papers and blogs on various topics (including AI, ML, and healthcare). He has been a speaker at various industry conferences, including the AWS Public Sector Summit, AWS re:Invent, and the American Public Human Services Association. He holds a master's degree in computer science from Temple University and completed a Chief Technology Officer program from the University of California, Berkeley.

Dr. Sherry Ding is an artificial intelligence and machine learning (AI/ML) technologist and evangelist with 20 years of experience in AI/ML research and applications. She currently works at Amazon Web Services as an AI/ML Specialist Solutions Architect, serving public sector customers on their AI/ML-related business challenges and guiding them to build highly reliable and scalable AI/ML applications on the cloud. Sherry holds a PhD in computer science from Korea University. She has authored more than 30 publications (including journal articles, book chapters, white papers, conference proceedings, and blogs) on different topics related to AI/ML. She is an active public speaker who has presented at various academia and industry conferences, such as IEEE conferences, AWS re:Invent, and AWS Summits.

About the Technical Reviewer

 **Krishnendu Dasgupta** is currently the Head of Machine Learning at Mondosano GmbH, leading data science initiatives focused on clinical trial recommendations and advanced patient health profiling through disease and drug data. Prior to this role, he cofounded DOCONVID AI, a startup that leveraged applied AI and medical imaging to detect lung abnormalities and neurological disorders.

With a strong background in computer science engineering, Krishnendu has more than a decade of experience in developing solutions and platforms using applied machine learning. His professional trajectory includes key positions at prestigious organizations such as NTT DATA, PwC, and Thoucentric.

Krishnendu's primary research interests include applied AI for graph machine learning, medical imaging, decentralized privacy-preserving machine learning in healthcare, applied artificial intelligence in robotics, and artificial intelligence computing. He also had the opportunity to participate in the esteemed Entrepreneurship and Innovation Bootcamp at the Massachusetts Institute of Technology, cohort of 2018 batch.

Beyond his professional endeavors, Krishnendu actively dedicates his time to research, collaborating with various research NGOs and universities worldwide. His focus is on applied AI and ML.

Acknowledgments

This book wouldn't have been possible without the contributions from many people.

We are grateful to our colleagues at Amazon Web Services and the representatives of various public sector organizations who engaged in ongoing discussions and shared their insights. These interactions were instrumental in shaping the core ideas presented throughout the book.

We extend our sincere thanks to Mickey Iqbal for taking the time to review our work and for crafting a thought-provoking foreword and Ben Snively for his insightful feedback and encouragement.

Our sincere appreciation to Mandy Kinne for her meticulous technical editing and creation of insightful diagrams. These elements significantly enhanced the clarity of the book.

Special thanks go to our technical reviewer, Krishnendu Dasgupta, whose constructive feedback helped us improve the quality of the book.

We are indebted to our editor and publisher for their patience and guidance throughout the entire process. Their expertise and support were essential in bringing this project to fruition.

Finally, our deepest appreciation goes to our families and loved ones for their unwavering love and support: to Soumya, Sravya, Sahiti, Shanti, Srinivas, and Toffee from Sanjeev; to Sandya, Suchir, and Sriya from Srinath; to Dianxun, Xizhen, and Albert from Sherry. (A special thanks to Suchir for his interest and for reviewing and providing feedback on the book.)

We would like to express our heartfelt gratitude to each and every one of you. Your contributions, big and small, were instrumental in making this book a reality.

Preface

Generative Artificial Intelligence (GenAI) has taken the world by storm. This powerful tool holds immense potential to improve efficiency, innovation, and constituent engagement within the public sector. From streamlining government services to enhancing public communication, GenAI offers a unique opportunity to address some of the most pressing challenges faced by government agencies, educational institutions, nonprofits, and healthcare organizations.

This book is your comprehensive guide to unlocking the potential of GenAI in the public sector. Whether you're a seasoned public sector professional, a policymaker curious about this new technology, or an IT professional seeking to expand your skill set, we'll equip you with the knowledge and practical steps to harness the power of GenAI for the benefit of constituents and communities. No prior experience with AI is required, as we'll build a strong foundation from the ground up.

The book is meticulously structured to guide you on your GenAI journey. Chapters 1–4 provide a solid understanding of AI fundamentals, introduce the concept of GenAI, and explore its relevance to the public sector. We'll then delve into developing a successful GenAI strategy, including navigating risks and challenges. Chapters 5–8 showcase practical applications of GenAI in various public service domains, from content generation to program management. Finally, we'll address implementation considerations, operation, maintenance, and emerging trends in GenAI technology.

While we recommend a sequential approach for foundational chapters (Chapters 1–4), the remaining chapters can be read independently based on your specific interests. The use cases and architectures presented throughout the book are illustrative, providing high-level concepts to spark your imagination and inspire your own GenAI projects.

Our ultimate goal is to empower you to navigate the world of GenAI with confidence, transforming public sector for the better.

Foreword

In a rapidly evolving world where technology and innovation shape the landscape of the public sector, the need for forward-thinking solutions and transformative ideas has never been more crucial. It is my pleasure to introduce *Empowering the Public Sector with Generative AI*.

Artificial intelligence (AI) has now seamlessly integrated into our daily lives and is rapidly reshaping our world with the potential to transform every facet of our society. Its influence spans from virtual assistants to autonomous vehicles, permeating nearly every aspect of our life and fundamentally altering the way we engage with the world around us.

Empowering the Public Sector with Generative AI stands at the forefront of this new era, offering a realistic perspective on how AI and generative technologies can modernize and transform public services and governance.

This book offers a practical outlook and prescriptive guidance on how GenAI can transform public governance, public sector customer experience, and service delivery – topics that are otherwise often overlooked in discussions about GenAI. This book serves as a valuable resource for public sector leaders, policymakers, technologists, and enthusiasts looking to understand how GenAI can revolutionize sectors such as healthcare, education, and state and local governments.

Whether it be automating administrative tasks or enhancing decision-making processes, GenAI holds the potential to streamline public services and enhance the well-being of citizens. The authors challenge us to reimagine the role of government in the digital age, inspiring us to think big and embrace the transformative potential of GenAI. They offer a comprehensive GenAI adoption strategy and a practical implementation road map for the seamless integration of this technology into public services, ensuring that its deployment is guided by ethical principles and an unwavering commitment to the welfare of citizens.

As we embark on this journey, it is vital to acknowledge the ethical and societal considerations that accompany any technological advancement. *Empowering the Public Sector with Generative AI* does not evade these crucial considerations; rather, it confronts them head-on. The authors underscore the significance of establishing robust ethical frameworks and cultivating public trust in the deployment of GenAI for the greater good.

As we navigate through complex challenges and unprecedented disruptions, the insights, use cases, and strategies presented in this book provide a road map for leveraging the power of GenAI to enhance efficiency, transparency, and citizen engagement. From smart governance to data-driven decision-making, *Empowering the Public Sector with Generative AI* illuminates the path toward a more inclusive and responsive public sector.

I am delighted to introduce this innovative publication, filled with thought-provoking contributions from experts and trendsetters in the fields of AI and public services. I trust this book will inspire and empower readers to embrace the possibilities of GenAI for creating a better future for all.

Mickey Iqbal
Managing Director of Public Sector Technologists,
Amazon Web Services

CHAPTER 1

Introduction to Generative AI

In the fall of 1989, the first year of our bachelor's degree program, a student asked the professor, "So, do you feed this flowchart to the computer and then it generates code?" The professor had this strange look and responded, "No, you need to develop the code, compile, and run it." Getting a computer to generate code using a flowchart was a silly thought or a brilliant thought at the time depending on how you look at it. Here we are, thirty or so years later: it is not only possible to generate code with simple natural language instructions but also possible to convert code from one language into another! And this is possible with Generative AI (GenAI)!

Artificial intelligence (AI) and machine learning (ML) have impacted our lives and virtually every industry in a meaningful way over the last decade. In fact, it is hard to fathom a life now without AI or ML, from Google Maps to Alexa to self-driving cars to Netflix recommendations and Amazon's excellence in delivering packages! The explosion of data, access to compute power, and advancements in ML algorithms have resulted in AI and ML being aspirational technologies to become integral parts of our lives and day-to-day business operations. AI and ML are revolutionizing enterprises by enhancing user experience, reducing costs, helping innovation, and improving operational efficiencies.

© Sanjeev Pulapaka, Srinath Godavarthi and Dr. Sherry Ding 2024
S. Pulapaka et al., *Empowering the Public Sector with Generative AI*,
https://doi.org/10.1007/979-8-8688-0473-1_1

GenAI goes one step ahead. We can now *generate* new content, such as contracting, marketing and sales documents, art, music and videos, architecture, design, and code, with natural language instructions. According to McKinsey, the potential total annual value of AI and analytics across various industry verticals is estimated at $9.5 trillion to $15.4 trillion,[1] and GenAI could add the equivalent of $2.6 trillion to $4.4 trillion annually, increasing the impact of AI by 15–40%.

So what exactly is GenAI? To understand it, we first need a conceptual grasp of what AI is; let us start by answering the question "What is AI?"

1.1 What Is AI?

AI has captivated the human imagination for a long time, for example, through movies such as *Star Wars* and sci-fi books including Isaac Asimov's talk about AI as a self-governing intelligence that has the ability to think independently and act like a human. But let us look at it from a more practical standpoint: we use AI in our daily lives in a number of ways.

Google Maps speaks to us in a similar way a human would, providing us with directions from one place to another. Amazon Alexa is a personal assistant that listens, performs actions, comprehends, and speaks naturally. Alexa answers verbal questions in a natural human speaking voice and can accomplish things such as checking the hours a store is open and calculating how long it'll take to drive there, along with performing a plethora of other tasks. Finally, there's Tesla. Tesla vehicles sense the surrounding environment – including traffic signals, other vehicles, and pedestrians – and adjust their actions accordingly. For example, a Tesla brakes automatically to avoid collisions.

---

[1] www.mckinsey.com/capabilities/mckinsey-digital/our-insights/the-economic-potential-of-generative-ai-the-next-productivity-frontier

To be more formal, in modern terms, AI is the ability of computer systems (also known as machines) to listen, comprehend, sense, and act like humans do. In other words, AI is the ability of machines to mimic human intelligence and cognitive functions.

If we continue to enhance and improve this technology and push AI concepts further to their natural conclusion, the result is a completely autonomous capability. An AI with autonomous capabilities is known as an Artificial General Intelligence (AGI). At this point in time, we haven't quite achieved AGI, but we're getting closer by the day.

1.2 AI and Estimating Home Prices

So, machines use AI to complete tasks without receiving explicit instructions in the form of programming logic. How can we use this capability in a practical way to, say, help estimate the price of a home?

Let's say you are in the Department of Housing and Urban Development. You need to estimate the sale value of a property in a particular neighborhood. You have a number of characteristics that describe the property, such as the square footage, number of bedrooms, number of bathrooms, and so on. We refer to these characteristics as *features*. Taking the features of a property into account, how can you calculate the price?

One way is to use a computer program with business logic, business rules, and calculations. You create a rules-based program to sum up the square footage of the house and multiply that with the estimated price per square foot based on the location. You also need to account for other features that typically influence the price of a house, for example, the age, the number of floors, the type of residence, neighboring amenities, and so on.

But how do you know how much each of these contributes to the overall price? How do you account for seasonal factors? What about flood zones? Or current economic conditions? Explicitly listing the value of each feature and factoring it into the actual house value is too complex to do accurately and is better solved using other techniques.

You can apply rules of thumb or *heuristic* methods to estimate the price, such as decrease $x amount for each year the house ages, increase $y amount for a brick front, decrease $z amount if it is in a flood zone. These heuristics are the contribution or *weight* that each of the features adds to the home price. Using this rules-based or heuristic technique, your computer can calculate the price of the house given the various feature weight inputs described earlier. This method is shown in Figure 1-1.

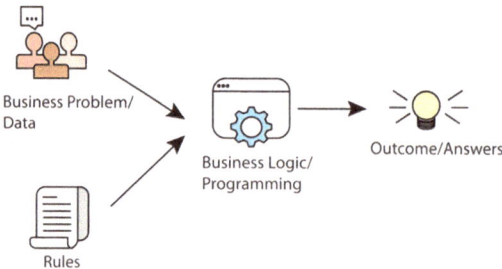

Figure 1-1. *Rules-Based Computer Programming*

However, using rules-based or heuristic programming to solve the problem raises several issues. One issue is generalization; for example, there's a chance that you haven't captured all the features that ultimately impact the final market value. There's also a high chance that the heuristics you use aren't accurate enough. Lastly, conditions may have changed and the rules that you use may not apply anymore; think here about the sudden change in market conditions in 2000, 2008, and during COVID in 2020 and how these factors might render heuristics too inaccurate to use.

Another method we can use to more accurately measure home price is a technique called machine learning (ML). This technique falls into the realm of modern AI.

1.2.1 Machine Learning

In machine learning, using the historical data and other patterns that you provide, the computer or machine comes up with a formula that can estimate the price of a house. ML is a two-step process shown in Figure 1-2: the first step is *training*, and the second is *prediction*, which is also sometimes called *inference*.

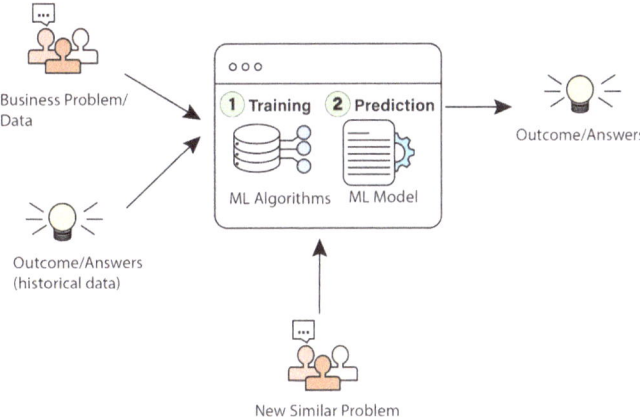

Figure 1-2. *Machine Learning*

In the first step, *training*, you provide a computer with historical data about various types of houses in your neighborhood, let's say the last twenty years of data. At a minimum, this data includes the price as well as various features of the house we talked about earlier. You also need to include other relevant features for which you may not have an accurate heuristic. For example, you may need to include an indicator that tells the computer that there is a movie theater or school near the house. The more data you provide under different conditions (such as economy, unemployment, interest rates), the better and easier it is for the computer to understand the pattern of the features. In addition to these features, you also provide the computer with the historical sale price for each of the houses; data with the feature combined with the outcome or the *label* is referred to as *labeled data*.

Note   This is called *supervised machine learning*; the computer or machine learns the patterns and characteristics that impact the value of the house.

In this case, using the historical data and other patterns that you provide, the computer creates a formula that can estimate the price of a house. You're not providing the business rules and calculations; the machine uses the historical data to come up with the formula and then optimizes the weights of each feature to get the overall value of the house. This equation, with all the optimized weights used in the prediction, is called an *ML model*.

The method by which the computer creates a model is called an *algorithm*. The type of algorithm used in this example is a simplified version of linear regression; there are several other types of algorithms including Logistic Regression, Support Vector Machines, Naive Bayes Algorithm, and Artificial Neural Networks (ANNs). Later, in this chapter, we'll explore ANNs.

Once you have the ML model, you can use this model to *predict* or infer the value of any house; this step in ML is also called the *inference*.

In the second step of our ML-based approach to estimate the value of a house, plug in all the features of the specific house you're trying to sell and the model provides an estimated price of that house. But how does this work? How does the algorithm create a model? Let's walk through it in more detail.

1.2.2 How Machine Learning Algorithms Work

First, the computer randomly assigns weights (the contribution amount) to each of the features you provided. So, for example, it randomly assigns 100.56 to the number of square feet in the house, assigns 10.08 to the number of floors, and so on for each feature; it is simply assigning a numeric decimal value to a feature. But what about those nonnumerical indicators, such as proximity to a school or movie theater? In that case, you assign a value of one to the feature if a school or movie theater is close or assign a value of zero for the absence of the feature.

Let's translate this into an equation:

```
V = (w0 x land value) + (w1 (total square footage)) +
(w2 (number of floors)) + (w3 (number of bedrooms)) +
(w4 (number of bathrooms)) + (w5x1)
```

where

V = the sale value
w(i) = the weight or contribution toward the sale value of each feature
(w5x1) = the presence of a school

Note that a bias term (w0) is added to manage the distributions of data. For example, in this case, the base land value may be considered as the bias. We don't need to really worry about it for the scope of this book. As we mentioned earlier, the computer assigns random weights w1, w2, and so on to each feature and comes up with a value (V).

The computer then compares that estimated value with the actual value of the house or similar houses. Since the weights were initialized at random, our value is probably wrong. The computer then changes the weights to determine a new value and so on; this continues until it finds weights that provide reasonably correct answers based on the values of the houses for which you provided data.

The weights are changed using optimization techniques such as Gradient Descent. A full discussion of Gradient Descent and other specific optimization techniques is outside the scope of this book; however, you can think of *Gradient Descent* as a means of continuously and iteratively adjusting the weights by minimizing the error or the difference between the actual and estimated values. In other words, Gradient Descent is used to optimize the model – to increase accuracy by minimizing the difference between predicted value and actual value of the house.

Sometimes, illustrating the final equation visually helps us better understand how it works. In order to do that with this example, we need to simplify the equation either as if 1) We were using one input – for example, the number of square feet (Figure 1-3) – or 2) As if we were using two inputs – the number of square feet and number of floors (Figure 1-4). Any more than two inputs implies a diagram with more than three dimensions, which is impossible for us to visualize.

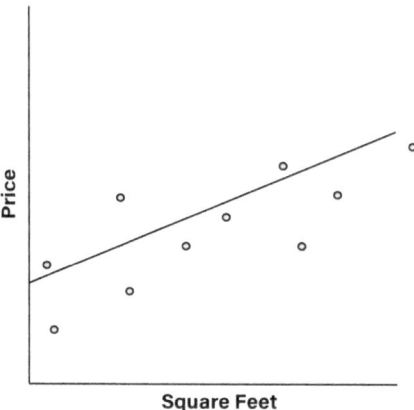

Figure 1-3. *Linear Regression with One Feature*

As Figure 1-3 shows, you can think of the equation or *model* as the straight line drawn on the graph. The straight line here provides the best approximation of the price of a house given the square feet. If we were to provide a new data point and add it to the graph, the square feet of another house, the model would determine the price by finding the corresponding value using the adjusted straight line.

The same concept extends for two inputs as shown in Figure 1-4. However, here the equation or model is represented by the floating green plane enclosed. You provide two inputs, and the model then gives you the corresponding price.

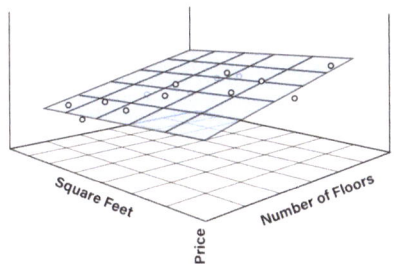

Figure 1-4. *Linear Regression in Two Dimensions*

We mentioned several types of algorithms, such as Logistic Regression, Support Vector Machines, Naive Bayes Algorithm, and Neural Networks. Each of these algorithms essentially does the same thing: they use an optimization technique to come up with an ML model that is then used to complete a task or make a prediction.

The task itself may vary – for example, the task was to estimate the price of a house. Another task is to look at a radiology image and determine whether or not there is a tumor or to take a conversation and determine whether the overall sentiment is positive, negative, or neutral.

You may be wondering how a machine is able to understand nonnumeric inputs such as images and sentences. You might already know that computers only work with numbers, particularly with binary digits or bits – zeros and ones. Machines cannot directly understand non-numeric inputs, but we can translate these types of inputs into numbers using certain techniques, which we'll cover later in the book when we look at the case studies.

Of these algorithms, Artificial Neural Networks (ANNs) or neural networks have become one of the most researched and used algorithms in ML, so much so that an entire branch of ML called deep learning is used to describe tasks that use these algorithms. Let's take a look at neural networks and deep learning.

1.2.3 Artificial Neural Networks and Deep Learning

At its core, an Artificial Neural Network is a machine learning algorithm. The term comes from initial efforts by researchers to mimic the way the human brain works. However, in reality, the two work quite differently, where the human brain is more complex than an Artificial Neural Network.

Let's take our same example problem to estimate or *predict* the house price. In a neural network, the features related to the property (such as the number of bedrooms, square footage, location, age, and so on) are fed into an abstract representation of an equation called a *neuron*. The equation works similarly to linear regression; each neuron receives multiple inputs related to each feature and creates an output.

Let's work through our property estimation example using one neuron, neuron 1. The equation for neuron 1 would be

$$\sum = w1^*x1 + w2^*x2 + w3^*x3 + w4^*x4 + \text{Bias}$$

where

x1 = square feet

x2 = number of floors

x3 = presence near a school

x4 = presence near a grocery store

$\sum$ = output

One difference between a neural network and linear regression is that we then convert this linear equation into a nonlinear equation using something called an *activation function* to make the model more flexible. The activation function is like using a dimmer switch for a light on the output of the equation. It also transforms the neuron into a decision-making system that can adjust responses to the data, helping the neural network learn and make better decisions.

To get an answer or *output* for our question, "What is the predicted sale price for a house?", we need another equation:

$$O = \sum (f)$$

where

O = output

$\sum$ = the results of the equation for neuron 1

f = the activation function

This equation is shown in Figure 1-5.

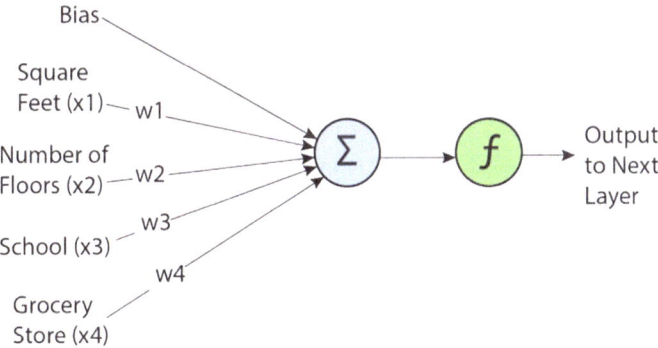

Figure 1-5. *Single Neuron for Predicting House Prices*

The actual activation function used varies depending on the type of learning task. Some examples include ReLU and Sigmoid, which takes a value and converts it to a value between zero and one.

Another difference between nonlinear and linear regression techniques is that each of the inputs is passed to a number of neurons, not just one. In other words, it is passed to a network of neurons stacked in multiple layers, hence the name *neural network*.

A neural network stacked in multiple layers is also called a *multilayer perceptron* (MLP). The MLP is a neural network divided into layers, with one layer as an input layer, another as an output layer, and one or many hidden layers in between, as shown in Figure 1-6. Every neuron in the network acts in the exact same way except for the type of activation function used for each layer. The type of activation function varies depending on the type of data and the type of output being generated.

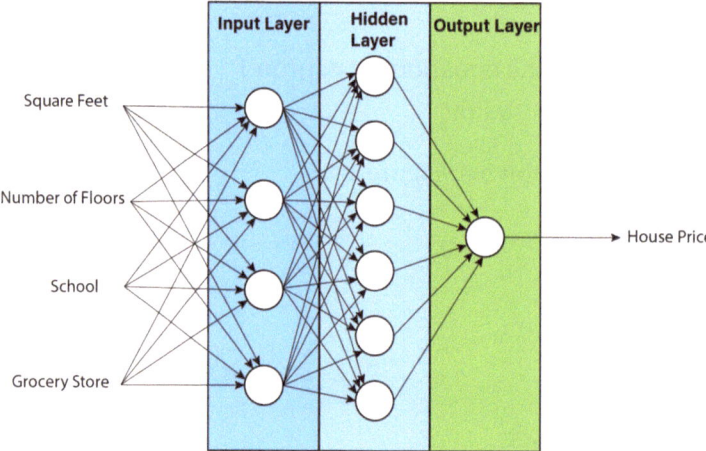

Figure 1-6. *Multilayer Perceptron (MLP) for Predicting House Prices*

The greater the number of layers is, the more complex the MLP is. In general, accomplishing tasks using a large number of layers is referred to as *deep learning*. The weights (values of w – a matrix of weights w1, w2, and so on) are adjusted or optimized in a way similar to the technique described for the linear regression. This ultimately results in a model that can be used on new data. The major difference is that the error from the output layer is used to derive the errors for each neuron using *backpropagation*, a mathematics formula that uses calculus.

You may be wondering why neural networks are necessary to understand GenAI.

Neural networks and deep learning are some of the fundamental forces driving the rise of AI. Self-driving cars, voice recognition devices such as Alexa, and many current AI-based technologies all use neural networks. Through the last decade, hardware and chips such as Graphical Processing Units (GPUs) have been invented to accelerate the processing of data for deep learning tasks. These developments have further propelled the use of deep learning into the realms of fields such as GenAI.

So far, we've focused on estimating the sale price of a house. This is just one type of machine learning or deep learning *task*. There are several different types of learning tasks that are important to understand these as well. Let's take a closer look.

1.3 Different Types of Learning Tasks

Machine learning or deep learning tasks fall into several broad categories. The task we've covered so far, estimating or predicting the price of a house, falls under something called *supervised learning*. Other types of tasks include unsupervised learning, reinforcement learning, and self- supervised learning. Let's review the concepts behind each of these.

1.3.1 Supervised Learning

As mentioned earlier, predicting house prices is a type of supervised learning task. In this ML type, the algorithm learns the mappings from input data to output predictions using a given dataset in the **training** phase. This type of input data is called "labeled" data where a label column (the actual value of the house) is provided together with all the other features. "Supervised" refers to the fact that the algorithm is guided by the labeled data to make predictions or decisions. The primary objective of supervised learning is to generalize from the training data, build an ML model, and then use that model to make accurate predictions on unseen, new data (inference).

Supervised learning can be further divided into two main categories:

1. **Classification**: In classification tasks, the goal is to assign input data points to predefined categories or classes. For example, for transportation applications, you pass in images of cars and trucks together with labels, build a model, and ask the model to detect whether a new image is a car or a truck. Common applications include spam email detection (spam or no spam), image classification (such as whether or not a radiology image contains a tumor), document classification, sentiment analysis, and so on.

2. **Regression**: In regression tasks, the algorithm predicts a continuous numerical value. The house price–based examples are in this category. Other applications could include forecasting budgets, forecasting demand of goods in a market, or estimating crop production.

1.3.2 Unsupervised Learning

Unsupervised learning differs from supervised learning in that it doesn't rely on labeled data for training. Instead, it focuses on uncovering patterns, structures, or relationships within the data without predefined target labels – unlabeled data.

One type of unsupervised learning task is clustering in which, for example, the constituents are divided into clusters with similar demographics during a census survey. Another type of unsupervised learning is dimensionality reduction. Let's say you have a dataset with a large number of features. Some of these features might be dependent on each other. In this case, adding these features to your model may not provide the best results. In dimensionality reduction, you use unsupervised learning to determine the features that are most representative or relevant to the dataset.

1.3.3 Reinforcement Learning

Reinforcement learning (RL) is a type of machine learning where a machine interacts with an environment and learns to make a sequence of decisions to maximize a reward signal. This machine is often called the "agent." The vacuum robots that clean your house are a classic example that use this method; RL relies on trial-and-error learning. For example, the robot vacuum initially learns your house layout by bumping into objects and taking note of each bump. Over time, the robot learns to take actions that minimize bumping.

RL is often used in scenarios where an agent needs to make a sequence of decisions over time, such as game playing, robotics, and autonomous navigation. Both the Mars Rover and Tesla are trained using RL for navigation. However, RL can also be used to fine-tune the performance of a machine learning model developed using other methods such as supervised or unsupervised learning.

1.3.4 Self-Supervised Learning

Self-supervised learning falls into the broader category of unsupervised learning. Both approaches aim to train a model from unlabeled data. In traditional unsupervised learning, the model doesn't have any specific task or labels to guide its learning; it typically focuses on clustering, dimensionality reduction, or density estimation. Self-supervised learning, on the other hand, involves creating supervised-like tasks within the unlabeled data itself. These tasks are specifically designed to capture meaningful information and relationships present in the data. One example of this is the Masked Language Model (MLM). In an MLM, random words in a sentence are replaced with a mask, and a neural network tries to predict the original word from the context provided by the other non-masked words in the sentence. For example, for the sentence "I want to apply for [MASK] benefits," the model might try to predict the word "healthcare" or "unemployment insurance" for the masked word.

1.4 What Is GenAI and Where Does It Fit?

Now that you have a basic understanding of neural networks and types of AI tasks, let's delve into what GenAI really is.

At its core, GenAI is a branch of deep learning that focuses on creating or generating new content using an input or a context. As such, it uses some of the same concepts we've already defined and described. The *task* for a GenAI model is to predict or generate content based on a given input. The content could be a word, an image, document, code, and so on, depending on what the task is.

Figure 1-7 illustrates the evolution of AI and GenAI since 1950.

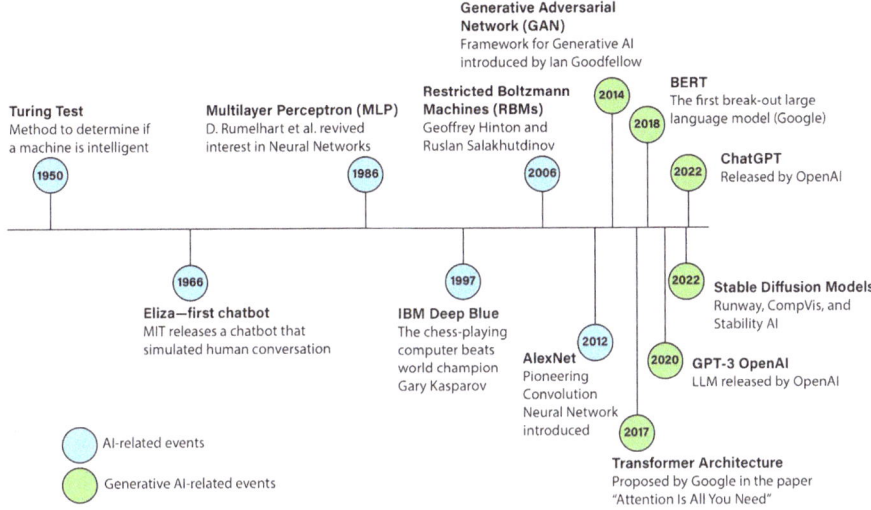

Figure 1-7. *AI and Generative AI Evolution Timeline*

We discussed the multilayer perceptron (MLP) earlier. MLPs were useful for pattern recognition or classification, but they struggled with any task that involved the generation of new content.

To solve this issue, additional variations of deep neural networks such as autoencoders, variational autoencoders (VAEs), and restricted Boltzmann machines (RBMs) were introduced.

Generative Adversarial Networks (GANs), introduced by Ian Goodfellow and his colleagues in 2014, marked a significant breakthrough in GenAI. GANs introduced a game-theoretic approach to training generative models.

In 2017, a new type of neural network called transformers[2] revolutionized natural language processing and generative modeling and ultimately led to the explosion in the scale and size of generative models such as BERT (Bidirectional Encoder Representations from Transformers) and GPT (Generative Pre-trained Transformer).

Transformers are complex neural networks especially attuned to processing large sets of input sequences of words in parallel. They use self-supervised learning during a pre-training phase to learn representations of language and context from unlabeled text data and, subsequently, need to be fine-tuned for specific supervised tasks.

Using transformers and hardware specifically developed for neural networks such as Graphical Processing Units (GPUs), we can train models on Internet-scale data. This is typically extremely large: think of the entire public datasets available on the Internet, for example, Wikipedia. The number of weights or parameters we mentioned earlier range in the order of millions, billions, and even trillions. This model is called a *Foundation Model (FM)*. An FM that takes words as inputs and generates words as outputs is called a *Large Language Model (LLM)* or a text-to-text generation model.

Once you have a pre-trained model, you can use it on a wide range of tasks. For example, an FM can be used to generate content on any given subject that can be found on the Internet. You can also use it in other ways, such as to summarize content for an agency-level executive briefing. We'll go deeper into the specifics on how an FM is trained later in this chapter.

An input to an FM is called a *prompt*, an instruction or a question, and the quality and type of output generated by an FM is largely dependent on the prompt. An entire field of creating these prompts has developed; we will cover *prompt engineering* in more detail in Chapter 4.

---

[2] https://proceedings.neurips.cc/paper_files/paper/2017/file/3f5ee24354 7dee91fbd053c1c4a845aa-Paper.pdf

An FM by itself is *stateless*. What does that mean? Once the final output is complete and the task is completed, the FM won't remember it for another task. If you want the model to remember the context of your previous question, you have to send that question as well as the response to that question again to the model. This is important for applications such as chatbots.

1.5 Examples of Foundation Models

There is no hiding the fact that the introduction of ChatGPT has led to the tremendous interest in GenAI. ChatGPT is a web application built to interact with an underlying LLM for natural language processing tasks. It was developed by OpenAI and launched in November 2022. The exact model used varies between different versions. For example, at the time of writing, we can use GPT 3.5, GPT 4, or GPT 4.0 Turbo, the latest version. GPT stands for Generative Pre-trained Transformer, which is a variation of the original transformer model.

We can also have FMs that can take in words as inputs and generate images as outputs. These are called *text-to-image generation models*. Stable Diffusion, introduced in August 2022, is an example of a text-to-image generation model.

In addition to text-to-text and text-to-image models, we can also have models that use multiple types of inputs such as images, text, audio, and video and can create multiple types of outputs! These are called *Multimodal Large Language Models* (MLLM). Multimodal stands for different modes, or different methods, by which information can be processed. Initial GPT models were text-to-text models, but the models have been evolving to include multimodal capability. We provide a more exhaustive list of FMs in Appendix A.

A detailed understanding about how FMs/LLMs work is useful in building and implementing GenAI applications. First, let's take a look at these concepts, and then we'll examine an image generation model such as Stable Diffusion.

1.6 How an LLM Works

At the core, an LLM is an FM that performs language-related tasks using a mix of unsupervised, self-supervised, and reinforcement learning techniques. These tasks might include predicting randomly masked words, helping understand context, determining sentence sequence, teaching coherence and flow, or reconstructing the original text from a compressed version focused on information preservation.

Similar to those in any other model, there are two steps in how the FM is able to do this: the first step is training, and the second is inference. However, for an FM, we label the first step as *pre-training*. The reason is that the model is first trained on a very large dataset of text and code, including books, articles, poems, and even code repositories. Later, it is optionally trained again for a specific dataset; this is called fine-tuning, which we will discuss in Chapter 4.

Now, let's take a deeper look at pre-training and inference.

1.6.1 Pre-training

To make it easier to understand what happens during pre-training, let's concentrate on the steps for one type of task: predicting randomly masked words.

1. **Data preparation**: The first step is to clean and process a massive dataset of text and code (books, articles, code repositories), breaking sentences down into tokens. A *token* is a word, subword, or

character. To be computed, the LLM has to turn each token into a numerical representation called an *embedding*, which captures meaning and relationships.

2. **Process embeddings**: The next step is to process the sequence of embeddings using Transformer layers. Each embedding examines all the other words in the input and calculates its relevance or how different words affect other words' meanings. This process is called *self-attention*, which enables the LLM to capture long-range dependencies and complex grammatical structures. The information flows through multiple layers that refine the LLM's understanding of word relationships.

3. **Output and fine-tuning**: Finally, after processing the embeddings, the LLM's main task is Masked Language Modeling (MLM), in which the LLM masks or replaces random words in the sentence with a special token. The LLM predicts the masked word based on the context of the remaining words. By predicting correct words repeatedly, the LLM learns associations between words and their contexts.

The task of selecting the best choice for the masked word isn't as simple as selecting the word with the most probable score; using this method, known as *greedy search*, tends to generate unimportant words, ignoring words that have the most contextual relevance. Instead, most models use some or all of the following techniques:

Beam search: This expands on greedy search by maintaining several promising word candidates (the beam size or beam width) at each step. It explores multiple paths and improves diversity and fluency compared to greedy search at the expense of increased computational cost.

Top-k sampling: Instead of directly picking the highest probability word, this method samples from a smaller set of the top K most probable words. This method introduces randomness and can lead to more surprising and creative outputs at the expense of generating occasional nonsensical sentences.

Temperature: This hyperparameter controls the randomness knob. A low temperature prioritizes high-probability words, leading to safer, more conservative outputs. A high temperature increases the influence of less likely words, encouraging exploration and potentially more novel, interesting results.

Top-p sampling or nucleus sampling: Instead of focusing on individual word probabilities, top-p sampling sums the probabilities of all the words. We set a threshold value, p, representing the desired proportion of the total probability mass we want to consider, then sort words by their individual probabilities in descending order. In this method, we keep adding words to the candidate pool until their cumulative probability reaches or exceeds p. Finally, we randomly choose the next word from this smaller pool of top-p candidates.

4. **Repeat**: Step 3, output- and fine-tuning, is repeated for every masked word in the dataset, iteratively improving the model's ability to predict the next word based on context.

5. **Feedback**: Some models incorporate this step; humans provide feedback to indicate the quality of the response (also known as Reinforcement Learning from Human Feedback or RLHF). This typically happens in a completely separate phase after the model has been through several rounds of steps 1–4. The model can then adjust its internal representations and probabilities to generate refined outputs that better align with user preferences or address identified issues.

1.6.2 Inference

Once the LLM has been pre-trained, it can now be used to generate words with a given input. To understand how this works, let's look at an example.

You may have heard this quote from Abraham Lincoln, who was a pivotal figure in US history: "…that government of the people, by the people, for the people, shall not perish from the earth."

Let's say, you forgot that quote and you need GenAI to reproduce it to see how the LLM would work. Note that this is a very simple example used for illustrative purposes; we really don't need GenAI for this when a simple Internet search would give us a correct answer.

Here is the sequence of steps that the LLM follows to generate the output for this task:

1. **Understand the prompt**: You provide an instruction to the model, "Complete the quote – that government of the people, by the people." This instruction is called a *prompt*. The model breaks it down into individual tokens and their embeddings, capturing their meaning and potential connections.

2. **Generate the next word**: Since the LLM has already been trained to predict a word given a set of words, it simply generates the next word. This involves using techniques such as beam search or top-k sampling to explore various possibilities of the word as described in the pre-training phase earlier in this chapter. In this case, it would be the word "For."

3. **Repeat**: The next word is added to the input sentence, and the rest of the words are again processed through the model. This iterative approach builds the entire quotation.

That's pretty much it; it is hard to imagine, but this is how an entire book can be written by the LLM, one word after the next. Of course, there is a lot more complexity and sophistication in the actual FMs that are out of scope of this book and not necessary for a basic understanding of GenAI.

We discussed how LLMs work; next, let's take a look at how image generation FMs work.

1.7 How Image Generation FMs Work

In LLMs, a model is pre-trained to estimate the next word given a sentence. As with LLMs, text-to-image models are pre-trained to generate images from text or a combination of text and images. Let's focus on one type of model called Stable Diffusion. The main architectural components of Stable Diffusion models are a variational autoencoder (VAE), forward and reverse diffusion, a noise predictor, and conditioning. Some components are used during pre-training, while others are used during inference (to generate an image from your prompt). Similar to our approach for an LLM, let's peel back the layers to examine what happens in each stage from a conceptual standpoint.

1.7.1 Pre-training

1. The encoding component of the variational autoencoder compresses an image into a smaller dimensional space called a *latent representation*. This reduces the amount of computational power required.

2. The model selects a random image and slowly transforms it into a noisy version in a series of steps called *forward diffusion*. Adding noise is equivalent to making an image blurry.

3. A U-Net, a specific type of neural network, learns to predict the noise of an image in each step. However, this process runs progressively backward, starting from the most noisy image to the cleanest image, and is called *reverse diffusion*. Research has shown that predicting noise in small incremental steps allows the model to learn the probabilities of data distribution much faster.

4. The model runs the preceding steps iteratively for billions of images until the U-Net learns to accurately predict noise in any image.

5. We train the model to generate images from text once the initial pre-training is done in steps 1–4. During this step, the model begins with random noise for the image. Then a prompt, in the form of text or text and image, is converted into a vector and injected into the U-Net as noise. The conversion of a prompt to a vector is handled by another model, such as Contrastive Language-Image Pre-training (CLIP), which is a neural network model based on the transformer architecture that can represent the relationship between text and images as vectors. The U-Net model then starts denoising the random vector using this noise as reference. The result after denoising is the latent representation of the desired image corresponding to the prompt entered. You can think of this whole process as starting with a chunk of clay (random noise), slowly molding it into an artifact (the latent image).

6. Step 5 is repeated for billions of prompts until the U-Net learns to accurately remove the noise generated by prompts.

The actual process of training is much more complicated than the simplified version provided earlier; there are many variations and options possible for training the U-Net. In addition, different types of loss functions as well as different parameters and equations are used for optimizing the model.

1.7.2 Inference

Now that we've pre-trained the model, typically using billions of images and prompts, we can use it to generate images. When you provide a prompt, the U-Net model constructs an image using the denoising process. Remember that this is the *latent image* or the representation of the image in a lower dimensional space.

During inference, the decoding component of the variational autoencoder recreates the final image in a higher dimensional space. The prompt was text or text and an image; if an image is provided, it influences the outcome of the final image.

The architecture of Stable Diffusion has been evolving, and the latest versions, such as Stable Diffusion 3.0, use transformers instead of the U-Net mentioned earlier. These models are beyond the scope of this book and are not really necessary for a basic understanding of GenAI.

1.8 Key Takeaways

In this section, we talked about how LLMs (text to text) and image generation (text to image) FMs work. If you found the concepts confusing, don't worry as these are advanced concepts in ML; there are only a few main takeaways to keep in mind; the rest are there to provide you with additional context and background information.

Let's examine the key takeaways from the previous section:

- A *Foundation Model (FM)* is a machine learning model that has been trained with a very large set of data.

- An FM takes an input known as a prompt to create an output. Every model has a limit on the size of the prompt (number of tokens).

- There are different types of FMs: text to text, text to image, and multimodal. Text-to-text models are also known as Large Language Models (LLMs). Multimodal models are often represented as Multimodal Large Language Models (MLLMs).

- An FM simply *generates* the next word or an image using an input sequence of words. Some people may think that the model is finding and using data from the Internet (as in an Internet search); that's not the case here. However, keep in mind that a front-end application, such as ChatGPT, that interfaces with an FM, such as GPT 3.5, may add features to incorporate Internet search results and then pass these search results to the FM as context. We'll talk about this in more detail in Chapter 4.

- Since the next word or image is generated purely based on probability, there is a chance that the model generates incorrect information. For example, in the previous scenario, the model could have generated "Of the people, by the people and for the Citizens." This is especially true if you modify the randomness or number of words available for selection using temperature, top k, or top p.

- The FM outputs can be modified by changing some parameters, such as temperature. In general, a higher temperature increases randomness, and a higher top k or top p increases the number of choices for words; both can increase the chance of incorrect words in the output. Therefore, it is important to carefully select the right values to give you the best results for your particular use case. For example, if you want the model to be creative for use cases such as poetry, you try increasing the temperature. Model cards, short documents that provide key information about the model, generally provide details on how you should set these values for the most desired results.

You should now understand some of the basic concepts behind GenAI. This groundwork will help you understand the use cases and applications discussed in the remainder of this book. As you read those chapters, you may often come back to this chapter to refresh your memory and get more insight into GenAI concepts.

1.9 Conclusion

In this chapter, we explored the fundamental concepts behind GenAI and its underlying technologies, such as machine learning, deep learning, and neural networks. We explored the inner workings of LLMs and text-to-image models, provided a conceptual overview of how to pre-train and use these models for inference tasks.

While some of the concepts may seem complex on your first read-through, the key takeaways are that FMs, similar to LLMs and text-to-image models, generate content based on prompts, using probability distributions learned from vast amounts of training data. GenAI has already made significant strides forward with models such as GPT, Claude, Gemini, and Stable Diffusion that enable us to generate new content, from text to images and beyond.

In the next chapter, we'll delve into the relevance of GenAI specifically in the public sector, exploring how this technology can be leveraged to address challenges and create value for constituents, employees, governments, and society as a whole.

CHAPTER 2

GenAI and the Public Sector

In Chapter 1, we discussed the fundamentals of Generative AI (GenAI) and provided a few examples of how it can be used in our day-to-day lives. This book, however, is primarily about the application of GenAI to the public sector. At its core, the mission of public sector organizations (PSOs) is to ensure the safety, well-being, and livelihood of the constituents they serve. However, delivering on this critical mission is an enormously complex undertaking given the scale, diversity of needs, regulatory requirements, and budgetary constraints involved. PSOs must grapple with providing seamless customer experiences, empowering employees, making data-driven decisions, optimizing operational costs, and safeguarding against threats, including fraud and cyber attacks.

This chapter explores how GenAI can serve as a powerful tool to help PSOs overcome these challenges and to fulfill their public service mandates. GenAI tasks spanning content generation, conversational interfaces, data analytics, summarization, and more can drive improvements in customer satisfaction, employee productivity, decision-making, and operational excellence.

Before we dive into understanding how GenAI can help PSOs, we first need to understand their key characteristics and the challenges that they face.

© Sanjeev Pulapaka, Srinath Godavarthi and Dr. Sherry Ding 2024
S. Pulapaka et al., *Empowering the Public Sector with Generative AI*,
https://doi.org/10.1007/979-8-8688-0473-1_2

2.1 Characteristics of a PSO

There are four key characteristics that distinguish a PSO from other commercial entities: mission, budget size, type and number of customers served, and employees. Let's take a closer look at each of these areas.

2.1.1 Mission

This characteristic perhaps describes the most important difference between a PSO and a commercial entity. Businesses prioritize maximizing profits and shareholder returns, while PSOs focus on providing essential services that ensure the safety, well-being, health, and livelihood of the public they serve. The specific goals of each PSO can vary depending on the sector they operate in, but all PSO goals are ultimately defined by their mission objective. Figure 2-1 illustrates PSOs across four different segments.

Government Agencies

- Improve citizen experience
- Improve agency staff productivity
- Improve business processes
- Improve contracting and procurement
- Minimize fraud, waste, and abuse

Service Providers, System Integrators

- Improve Software development processes – architecture, design, code generation
- Improve program and project management
- Deliver RFX response
- Improve employee productivity

Non-Profit/Healthcare Organizations

- Improve paient outcomes in healthcare settings
- Improve provider productivity/medical diagnoses
- Improve disaster relief, humanitarian aid

Educational Institutions

- Improve STEM education—content generation
- Personalized lesson plans
- Language translation and equitable learning
- Improve instructor productivity

Figure 2-1. *Public Sector Organization Segments and Focus Areas*

As you can see from Figure 2-1, PSOs typically fall into the following categories:

- **Government agencies**: Provide critical services to the constituents, including homeland security, healthcare (Medicare, Medicaid, children's health insurance), justice and public safety, social and safety net services (social security, unemployment insurance, food stamps, child welfare, etc.)

- **Service providers and system integrators**: Partner closely with the government agencies and provide them with products and services to help achieve the mission objectives

- **Nonprofit government and healthcare organizations**:
  Work directly with the government to provide services
  in case of disasters or other times of need

- **Educational institutions**: Provide education to the
  constituents

2.1.2 Budget Size

PSOs are generally allocated a large portion of the government's overall
budget spend as the needs for PSOs can be significant. Figure 2-2 provides
an overview of the spend in the US public sector in 2023.

data source: "The Budget and Economic Outlook: 2024 to 2034" February 7, 2024,
https://www.cbo.gov/publication/59710

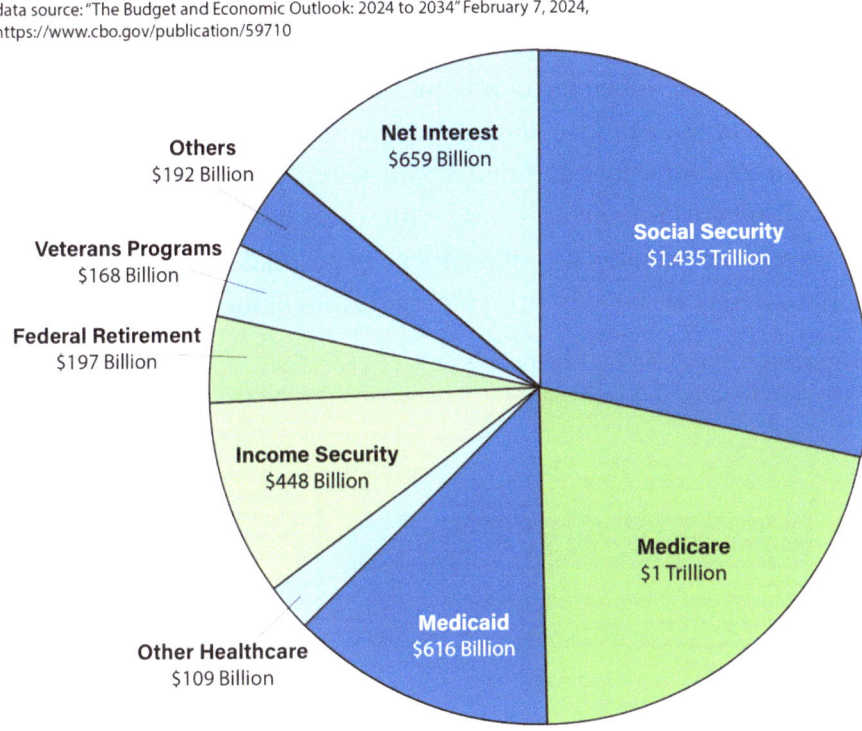

Figure 2-2. *2023 Actual US Federal Government Spending in the
Public Sector*

2.1.3 Type and Number of Customers Served

The impact of the public sector is immense. It is felt by nearly every constituent at some time or another in their lifetime. PSOs serve tens of thousands of people for education organizations to hundreds of millions of people for government agencies.[1] While every constituent depends on the government in some way or another, Figure 2-3 depicts the number of people that receive benefits in certain categories in the United States.

data source: "Policy Basics: Where Do Our Tax Dollars Go?"
https://www.cbpp.org/research/policy-basics-where-do-our-federal-tax-dollars-go

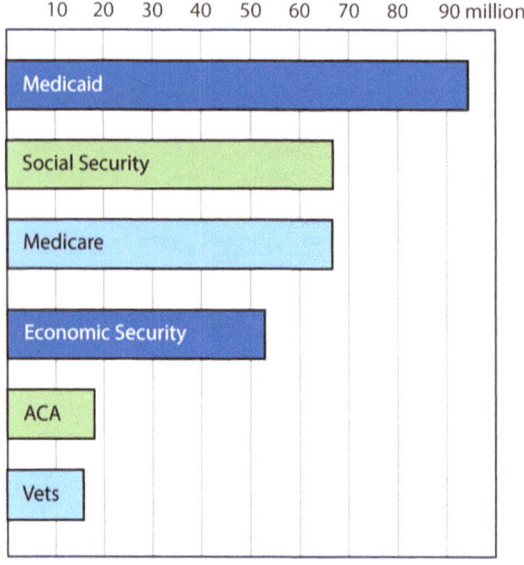

Figure 2-3. *Number of Beneficiaries by Program*

---

[1] www.cbpp.org/research/policy-basics-where-do-our-federal-tax-dollars-go

2.1.4 Employees

PSOs employ tens of thousands to hundreds of thousands people. The US public sector is also one of the largest employers in the country; as of 2023, the federal government (civilian) employed 2.95 million people.[2] These include both the agency staff that are constituent-facing as well as the leaders and administrators that deal with program management and agency administration. This workforce is largely responsible for how the government functions, and they have the critical mission of ensuring that the level of service provided meets the needs and expectations of millions of constituents.

2.2 Public Sector Challenges: A Closer Look

As described earlier, the public sector is vast and complex and consequently faces several challenges. Let's take a closer look at the challenges PSOs face in their daily operations. Let's explore these challenges through the example of a PSO providing safety net services, such as unemployment insurance and food stamps.

2.2.1 Customer and Employee Experience

Let's consider Jane, a single mother of two kids who lost her job during the pandemic. Jane struggles to feed her family and decides to apply for unemployment insurance and food stamps. She encounters a manual benefits application process which takes weeks to complete. With immediate food needs, she relies on charities while navigating long wait times for status updates on her benefits application. This is an indicator of poor constituent experience that ultimately leads to the PSO losing trust with constituents.

---

[2] https://usafacts.org/articles/how-many-people-work-for-the-federal-government/

Similarly, the agency caseworker John faces a daily backlog of applications due to outdated workflows and lack of process automation for benefits eligibility determination. He wants to help Jane quickly but is hindered by slow processing and manual tasks, including checking policies to verify qualification for benefits. This indicates poor employee experience.

2.2.2 Data-Driven Decision-Making and Operational Costs

Benefits administrator Xiulan requires better insights into the benefits program operations to take data-driven decisions. She is also concerned about meeting compliance and regulatory requirements. Reports on performance metrics and application processing times are crucial, but for generating them, she is dependent on the IT team. She strives to optimize costs, yet manual processes and legacy systems drain the budget. Balancing the legacy system transformation with budget constraints adds another layer of complexity. Additionally, Xiulan is constantly under pressure to meet regulatory requirements.

2.2.3 IT Systems and Technical Debt

Mary, the IT director, collaborates with Xiulan to provide reports. However, outdated IT systems and complex Software Development Life Cycles (SDLCs) consume significant time and resources within her department. Minimizing this burden would allow her to implement strategic initiatives, a goal shared by CIO Maria.

2.2.4 **Cybersecurity Concerns**

The Chief Information Security Officer Arnav is tied up with constant cybersecurity threats and vulnerabilities while safeguarding constituent data within the same legacy systems. This pressure to protect sensitive information with outdated technology presents a significant challenge.

These stories illustrate the multifaceted challenges PSOs face: delivering effective and efficient services under tight budget constraints.

2.2.5 **Other Challenges**

In addition to the preceding challenges, let's look at other challenges that impact PSOs:

- **Fraud, waste, and abuse (FWA)**: PSOs must proactively detect and prevent FWA to minimize its impact. For instance, according to the US Government Accountability Office,[3] the estimated unemployment insurance fraud during the pandemic was $100 billion to $135 billion, highlighting the need for robust measures.

- **Document overload**: The US federal government alone manages over 9800 unique forms costing the public over $100 billion on paper processing.[4] This, combined with legacy systems and manual workflows, signifies inefficient and time-consuming processes within many PSOs.

---

[3] www.gao.gov/products/gao-23-106696

[4] www.federaltimes.com/federal-oversight/2022/10/20/filling-out-10000-government-forms-cost-public-billions-chamber-says/

- **Contracting processes**: Updating legacy IT systems often involves lengthy contracting procedures within the federal government. RFP (request for proposal) preparation, analysis, and subsequent delays contribute to a prolonged procurement cycle. According to a Gartner survey,[5] with an average of 22 months, the public sector has the longest buying cycles for technology purchases compared to other industries.

- **Serving a diverse population**: According to the US Census Bureau,[6] people in the United States alone communicate in more than 350 languages. The significant increase in non-English speakers necessitates the availability of communication materials, website content, and public documents in multiple languages, eliminating the need for additional translators.

2.3 How Can GenAI Help PSOs?

GenAI has the potential to address many of the preceding challenges, enabling PSOs to meet their mission and implement their strategic goals and objectives. Let's explore how.

---

[5] www.gartner.com/en/newsroom/press-releases/2022-09-06-gartner-survey-finds-government-tech-purchase-decisio

[6] www.usa.gov/official-language-of-us

2.3.1 Content Generation

In Chapter 1, we discussed how GenAI can create new content using an input. Therefore, you can use it to develop different types of content which PSOs deal with on a regular basis, such as strategy documents, various contract-related material such as request for proposals (RFPs), and even program management reports. This dramatically improves productivity because much of the content can be reused.

PSOs face significant challenges with legacy and antiquated systems and legacy software codebase and undertake digital transformation initiatives. GenAI can help with Software Development Life Cycle (SDLC) tasks. With GenAI, technical teams can generate code and perform related tasks, such as securing and debugging code. They can also generate test plans, test code, draft architecture, and design documents, among other artifacts. We'll cover content generation, including document, image, and code generation, in extensive detail in Chapter 5.

2.3.2 Conversational Agents or Chatbots

GenAI can retrieve relevant content that helps constituents and staff. For example, rather than searching through a 500-page policy document for specific guidance, John, the caseworker, can ask a GenAI-powered chatbot his questions and get answers almost instantaneously.

Some might point out that such bots exist currently: What's the difference between those bots and a GenAI-powered bot?

At a high level, when you ask a GenAI-powered bot a question, it uses a Foundation Model (FM) that takes in the question as the input or context, and it *generates* the output. When looking for relevant material, that material is also passed in as context. This concept is *Retrieval-Augmented Generation (RAG)*, and we'll dive deeper into it in Chapter 4.

The difference between this technique and traditional bots is that with traditional bots, you program the bot with a number of potential static canned input questions and the corresponding output answers. With GenAI, there is no need to provide the output first. As a result, the type of questions GenAI can handle increases dramatically compared to traditional chatbots.

You can use the same techniques for external websites and applications. Now, Jane, the single mother looking for benefits, can easily access the information she wants on her application online, and John, the caseworker, can spend more time on complex cases that require human intervention. Jane can also converse with a chat or using multiple languages if needed. We'll cover chatbots extensively in Chapter 6.

2.3.3 Content Summarization

GenAI can be used to summarize documents, case notes, contact center call transcripts, and other content to generate an executive summary. For this technique, you feed in the data that you need summarized into an FM which generates the summary. This is particularly useful in customer service center contexts where the contact center agent can retrieve a summary immediately during or after a customer support call instead of spending a lot of time sifting through the case history. We'll explore this technique more later in Chapter 7.

2.3.4 Business Intelligence, Analytics, and Reporting

Currently, many government agencies use some kind of business intelligence platform or analytic tools that use Structured Query Language (SQL) for analysis and reporting. The technical nature of SQL sometimes hinders the adoption and use of these tools across different types of non-technical personas within a PSOs. GenAI can empower program leaders and agency administrators by enabling them to use natural language for analytics and reporting. This minimizes a PSO's dependency on technical teams and reduces delays in getting data-driven insights. This is very useful to PSO administrators, management, and executive teams. GenAI can also help security teams to analyze data for security-related issues. We'll cover program management, business intelligence, analytics, and reporting extensively in Chapter 8.

The preceding tasks show the potential for GenAI in the public sector. It is no surprise that a number of PSOs have already started to actively explore GenAI use cases in the United States.[7]

2.4 Conclusion

The examples in this chapter are just a few of the tasks and efficiencies that GenAI can introduce into a PSO. We'll cover these as well as additional tasks in much more detail in Chapters 5–8. Even at this high level, we can already see some of the distinct advantages for PSOs to use GenAI: improved productivity and efficiency, improved customer experience, and cost/budget optimization. The overall outcome is efficient and effective public sector services that positively impact the well-being and livelihood of constituents.

However, as we've discussed, the public sector's mission is critical, impacting the lives of hundreds of millions of constituents, so it is essential to approach GenAI adoption with a clear strategy and a well-defined blueprint. In the next chapter, we'll explore approaches to determine whether or not GenAI is the right choice for your agency, define what a successful implementation blueprint looks like, and identify best practices for mitigating potential risks and pitfalls. By addressing each of these critical aspects, you'll be better equipped to harness the power of GenAI while safeguarding your organization's integrity and upholding the public trust.

---

[7] www.dhs.gov/data/AI_inventory

CHAPTER 3

GenAI Strategy: A Blueprint for Successful Adoption

Maria, the Chief Information Officer of a public sector organization (PSO), is excited with the promise and potential of GenAI (GenAI). Her organization is struggling with poor constituent experience and low workforce productivity. The PSO seeks improvements in both these areas. Maria has heard from various industry experts that her organization should consider using GenAI to address some of these challenges.

However, Maria has a number of questions and concerns about GenAI including

1. What should be my GenAI strategy and what does a blueprint look like?

2. Where should I start and how do I implement GenAI within my organization?

3. What are the major risks and pitfalls of GenAI?

4. How can I overcome these risks and pitfalls? What are the GenAI implementation best practices?

© Sanjeev Pulapaka, Srinath Godavarthi and Dr. Sherry Ding 2024
S. Pulapaka et al., *Empowering the Public Sector with Generative AI*,
https://doi.org/10.1007/979-8-8688-0473-1_3

These are the specific issues that we will discuss in this chapter as we lay out a high-level GenAI strategy and blueprint, a methodology for implementation, and a framework to address various challenges that agencies face during implementation.

3.1 GenAI Strategy and Blueprint

Let's first tackle Maria's question about strategy and blueprint. At a high level, a GenAI strategy ensures that GenAI implementations align with overall mission objectives to help constituents as well as the PSO workforce. Figure 3-1 provides a blueprint that PSOs can use to develop this strategy.

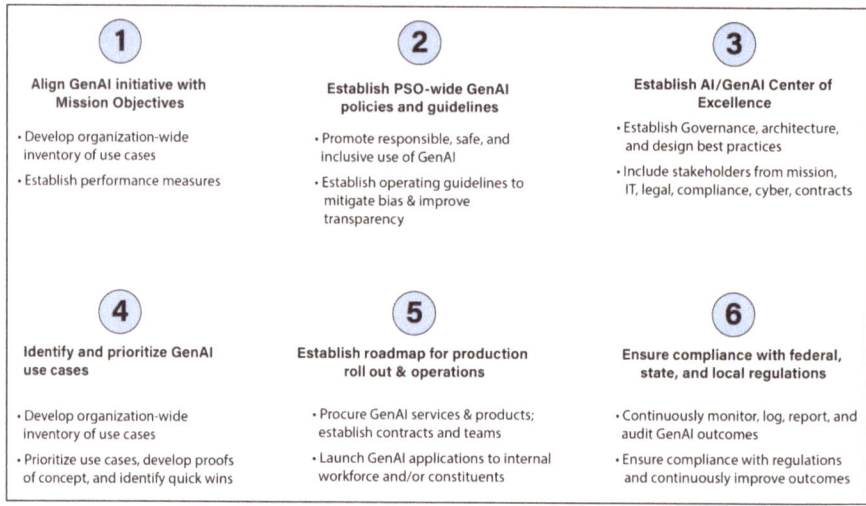

Figure 3-1. *GenAI Strategy Blueprint at the Organization Level for Public Sector Organizations*

Let's review each of the major strategy areas of the blueprint.

3.1.1 Align GenAI with Mission Objectives

Ensure that any GenAI initiative undertaken is fully aligned with mission objectives and strategic goals of the organization. Otherwise, the initiatives will struggle to provide value. Additionally, develop an outcome-based strategy for GenAI, establishing performance measures to evaluate the effectiveness of the GenAI program. As an example, outcomes include improvements in constituent experience, satisfaction, employee productivity gains, and time to service delivery, among others.

3.1.2 Establish PSO-Wide Policies, Acquisition, and Operating Guidelines

PSOs have a unique responsibility to prioritize constituent safety and security when implementing any new technology, including GenAI. To ensure responsible, safe, reliable, and secure use of GenAI, establish specific organization-level policies, acquisition, and operating guidelines that go beyond standard technology protocols. These should address key areas such as transparency and explainability, bias and fairness, data privacy, and security. Also, establish clear lines of responsibility for the development, deployment, and monitoring of GenAI systems. Evaluate current contract and procurement practices for any changes needed to acquire GenAI services. Additional budget may be needed to upgrade and modernize data platforms and/or legacy environments.

3.1.3 Establish AI/GenAI Center of Excellence (ACOE)

This is critical to provide organization-level guidance on GenAI governance, architecture, responsible use, and implementation best practices. This ACOE can include stakeholders from various teams including the mission, legal, compliance, cyber, contracts, and IT.

3.1.4 Identify and Prioritize GenAI Use Cases

Maintain an organization-wide inventory of use cases, prioritized based on mission needs and value delivered. Identify quick wins to experiment with a proof of concept and prove value to the mission. Be prepared to fail fast and apply lessons learned. This approach can help the agency pivot to choosing the right set of use cases and/or change the tactical plans in case of failures.

3.1.5 Establish Tactical Road Map for Production Rollout and Operations

Based on the priority use cases, develop and implement a proof of concept application (POC). Validate the outcomes with mission stakeholders to get alignment on implementation. Once a POC is complete, start full-fledged application development and plan for the production rollout to constituents (external users) and the agency workforce (internal users). Start with internal users first to mitigate risks before rolling out to constituents. At each state of the process, ensure alignment with the established performance measures/outcomes and adjust the plan accordingly.

3.1.6 Ensure Compliance with Federal, State, and Local GenAI and AI Guidance and Regulations

Note that there are a number of regulations and policies at the federal, state, and local levels. As an example, the *Bipartisan Framework for US AI Act*[1] provides general guidance on AI, including defending national security, protecting consumers and kids, and promoting transparency.

---

[1] www.blumenthal.senate.gov/imo/media/doc/09072023bipartisanaifra
mework.pdf

The *NIST Artificial Intelligence Risk Management Framework: Generative Artificial Intelligence Profile*[2] provides guidance to improve the ability of organizations to incorporate trustworthiness considerations into the design, development, use, and evaluation of AI products, services, and systems. Additionally, the *Blueprint for an AI Bill of Rights*[3] from the White House Office of Science and Technology Policy is a set of five principles to help guide, use, and deploy systems in the age of AI. Many states have enacted bills for regulating the use of AI,[4] and cities have also enacted laws.[5] Based on the jurisdiction, ensure compliance with these regulations by continuously evaluating, tracking, auditing, and reporting on your GenAI operational outcomes.

3.2 GenAI Implementation

The blueprint described earlier helps PSOs with developing the strategy necessary to adopt GenAI in a thoughtful and consistent manner, ensuring alignment with the overall mission of a PSO. However, we also need to answer Maria's question: Where should I start and how do I implement GenAI within my organization? For this, a strategy alone is not enough. What is needed is a methodology to effectively manage projects for GenAI implementation. Figure 3-2 illustrates this methodology.

---

[2] https://airc.nist.gov/docs/NIST.AI.600-1.GenAI-Profile.ipd.pdf

[3] www.whitehouse.gov/ostp/ai-bill-of-rights/

[4] www.csg.org/2023/12/06/artificial-intelligence-in-the-states-emerging-legislation/

[5] www.nyc.gov/assets/dca/downloads/pdf/about/DCWP-AEDT-FAQ.pdf

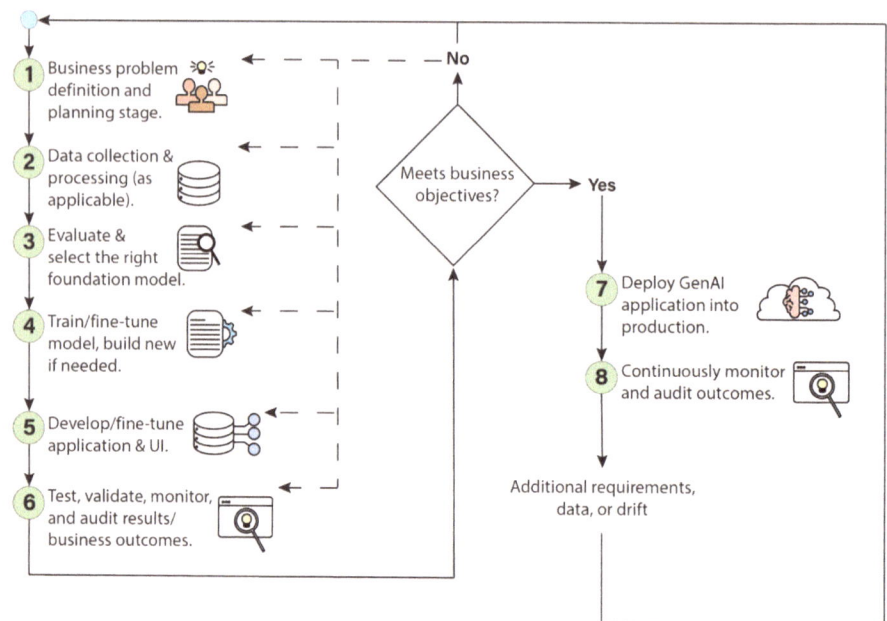

Figure 3-2. *GenAI Life Cycle*

As Figure 3-2 shows, the methodology consists of eight different phases that form a cycle, which we call the GenAI life cycle. Let's discuss each phase of the life cycle in detail.

3.2.1 Business Problem Definition and Planning Stage

Keeping the mission and customer in mind and working backward is a great way to start GenAI initiatives. In this phase, identify who the customers are; what their problems, challenges, and pain points are; and how GenAI can help solve these problems. These customers can be external users, such as constituents, or internal users, such as the agency workforce. As part of this effort, identify the specific use cases that align with your mission objectives. These could include improving constituent experience, improving employee experience, improving decision-making, predicting trends, enhancing communication, or personalizing interactions. Once you identify the problem and the use cases, evaluate whether using GenAI truly solves the problem. Sometimes, there is a tendency to assume emerging technologies can solve all problems. Undertake an assessment and validate with a proof of concept on whether the problem can be solved with GenAI before implementing a solution.

3.2.2 Data Collection and Processing

Data is the foundation for GenAI outcomes. Therefore, plan the data strategy, in other words, how and where to get, store, and manage relevant data to use with Foundation Models (FMs) depending on the use case.

3.2.3 FM Evaluation and Selection

In this phase, the first step is to assess the approach to procuring the FMs. We discussed FMs in Chapter 1. In general, there are three options: (a) use an existing model with no other modifications, (b) fine-tune the model, or (c) develop a custom model. The decision comes down to the requirements for your specific business case and available budgets. If you choose to directly use or fine-tune an existing FM, selecting the right model for the use case is a critical step. To select the best model, consider a number of factors, including the level of customization, inference options, licensing agreements, and pre-trained model behavior (that includes context windows and latency). We will discuss these concepts in more detail later in this chapter.

3.2.4 FM Training and Fine-Tuning

In this phase, based on the option selected for FMs, either build or fine-tune the FM. Note that building a custom model is a complex undertaking because you need to consider various factors, including the approach,

51

data, time, and costs (labor and infrastructure). If a pre-trained model does not require any fine-tuning, then this phase is not needed.

3.2.5 Application and Orchestration Layer Development

Once you decide on the FM, you need an application layer to interact with end users' inputs or prompts, the data, and the FMs. You also need an orchestration layer to integrate with existing systems and manage various steps required for interacting with FMs. Providing this layer involves developing APIs, connectors, or interfaces that allow seamless interaction with various systems to serve end-user requests. It may also include developing user-friendly interfaces for constituents and agency staff. We will discuss all this in more detail in Chapter 4.

3.2.6 Testing, Validation, Monitoring, and Auditing

This phase involves establishing processes and metrics to thoroughly test, validate, monitor, and audit the results, to ensure alignment with the mission and business outcomes. This is especially critical since FMs have a risk of producing hallucinations, whereby the models produce false content as if it is accurate. Based on the results, you would regularly update and fine-tune the models. This is to improve accuracy and adapt to changing trends. Also, continuously audit the models to make sure that they meet business outcomes. We will cover some of the techniques to reduce hallucinations in Chapter 4.

3.2.7 Production Deployment

If the initial outcomes are satisfactory, plan for production deployment of the GenAI application. Note that you would have to consider a number of factors including performance, scalability, availability, disaster recovery, and so on, as well as consider costs and trade-offs with these factors. Each of these factors has a great impact on the end-user experience. We will cover these in more detail in Chapter 9.

3.2.8 Continuous Monitoring, Auditing, and Fine-Tuning

Once the GenAI application is deployed into production, it is critical to continuously monitor and audit the application to ensure that it is meeting the mission outcomes. It is also important to ensure that the application is operating in compliance with federal regulations. If there is any deviation or drift, the application and/or the FMs need to be adjusted.

The GenAI life cycle captures important steps in implementing GenAI solutions. However, it is critical to keep in mind that responsible AI practices should be adopted throughout the life cycle. Practicing *responsible AI* use includes minimizing bias, improving transparency, accountability, and fairness. Responsible AI also includes contemplating and mitigating the ethical concerns that are part of its development, deployment, and impact on the constituents and society. We talk about this in more detail later in this chapter.

3.3 GenAI Risks and Challenges Within the Public Sector

As we outlined in Chapter 2, GenAI can have a positive impact on the public sector in the areas of constituent experience and staff productivity, among others. However, what about Maria's question on the risks and challenges PSOs face with a GenAI implementation? Indeed, there are inherent risks specifically with GenAI applications and with any AI technology in general. Some of those risks and challenges include

- Data bias

- Data privacy and security concerns

- Content safety: misinformation and disinformation

- Lack of transparency and explainability

- Social and economic impact

- Model bias and discrimination

- Regulatory compliance, legal, copyright, and liability

- Challenges with people, process, technology, and data

Let's review each of the preceding risks and challenges in more detail.

3.3.1 Data Bias

FMs are trained on massive datasets of text and code, which can contain biases and prejudices. This can lead to the model generating output that is discriminatory, unfair, or offensive. For example, a model trained on a dataset of news articles may perpetuate gender stereotypes in its writing.

3.3.2 Data Privacy and Security

For some use cases, FMs may have access to sensitive personal data, such as names, addresses, and financial information.

This data could be unintentionally leaked or hacked, leading to identity theft or other forms of harm. Also, since GenAI depends heavily on the input prompt and context, any manipulation of the input can result in adverse consequences. People with malicious intentions can hack your prompt and alter it. People could also find ways to get around some of the safeguards developed in models and get answers to questions containing information they shouldn't have access to.

3.3.3 Content Safety: Misinformation and Disinformation

The content created by GenAI can be used for activities that are detrimental to society. For example, a generated video could make it appear as if a PSO official said something that they never said in reality. GenAI can be used to create fake news articles that are indistinguishable from real news. This can lead to confusion and the erosion of trust in legitimate news sources.

3.3.4 Lack of Transparency and Explainability

It is often difficult to understand how GenAI models make decisions, which can lead to concerns about fairness and accountability. For example, it may be difficult to determine why a model flagged a social media post as harmful by a security agency. Malicious actors could potentially exploit the lack of transparency in GenAI models to manipulate them into generating harmful outputs, such as generating hate speech or propaganda.

3.3.5 Social and Economic Impact

As part of the workforce fears, GenAI has the potential to automate many tasks currently performed by humans, leading to job displacement in certain industries. This could exacerbate inequality and create social unrest.

3.3.6 Model Bias and Discrimination

FMs can perpetuate and amplify existing social biases, leading to discrimination against certain groups of people. For example, a model used to make hiring decisions could inadvertently discriminate against females or minorities.

3.3.7 Regulatory Compliance, Legal, Copyright, and Liability

GenAI solutions that are used to create content including text and images can pose several compliance, legal, copyright, and liability issues. Let's examine them closely.

PSOs are subjected to a number of federal, state, and local government regulations. GenAI applications including those that generate content, summarize content, or provide services to constituents such as chatbots must comply with these regulations. Any violations can lead to loss of funding and/or other penalties.

One of the major issues with GenAI is identifying the ownership – for example, let's say that a training coordinator within a PSO creates images from original content; the copyright may belong to the original content creator. This poses challenges especially in the public sector. Additionally, GenAI models are trained on large datasets that may include copyright material, and hence the generated content poses the risk of copyright infringement. GenAI solutions can also lead to liability challenges especially when a GenAI-generated content results in harm to the constituents.

3.3.8 Challenges with People, Process, Technology, and Data

As we discussed earlier, PSOs are subject to strict security, regulatory, and compliance requirements. They face increased scrutiny due to their interaction with the public. As a result, they tend to be more risk-averse with processes that may often be time-consuming with multiple stages of the GenAI life cycle. Let's take a look at some of the following challenges. We subsequently present a framework that can help PSOs deal with these risks and challenges for successful GenAI adoption. The challenges are best considered from four broad perspectives: people, process, technology, and data.

People

Procuring and implementing GenAI in the public sector presents a unique set of people-related and cultural challenges that agencies need to address. One of the most difficult challenges is to alleviate the concerns of employees who think GenAI is a potential replacement for them in their jobs, which brings adoption resistance. As such, educating employees, managing change, and addressing employee concerns are crucial.

Process

GenAI is a rapidly evolving field with changes almost every day. In addition, it is also relatively new with a number of unknowns and potential risks as outlined earlier. PSOs tend to be risk-averse and rightly so, given that the stakeholders are the public, and any GenAI outcomes that are not accurate can have a serious impact on the livelihood and well-being of the constituents. It is critical to develop proper governance and process structure to address these concerns. In addition, it is also critical to ensure that responsible AI practices are included in the GenAI implementation processes.

Technology

Acquiring, implementing, and maintaining GenAI systems may require significant computational resources, including powerful hardware and storage capacities, as well as various types of software and development tools. PSOs must plan for these resource requirements.

Data

Data is the foundation for GenAI, and higher data quality leads to better GenAI models/applications with better business outcomes. Having a concrete data strategy is key to a successful GenAI adoption. Acquiring the necessary data for customizing, fine-tuning, and deploying GenAI applications can be a time-consuming process. PSOs must navigate data acquisition, data sharing agreements, and data quality issues. We will discuss the importance of data extensively later in this chapter.

Addressing the preceding challenges requires careful planning, collaboration with experts in AI ethics and regulation, and the development of clear guidelines and policies specific to GenAI adoption in the public sector. Agencies must prioritize security, ethical considerations, transparency, and content control to ensure that GenAI is used responsibly and effectively.

3.4 High-Level GenAI Implementation Framework

You may be wondering whether it is really worth implementing GenAI given all the risks and challenges we discussed so far. As we argued in Chapter 2, given the tremendous benefits offered by GenAI, we say yes, as long as you carefully consider and implement steps to address these risks and challenges. This also brings us to Maria's questions: How can I overcome the risks and pitfalls with a GenAI adoption? What are the best practices? To address these questions, we present a high-level framework using the four perspectives described in the previous section: people, process, technology, and data.

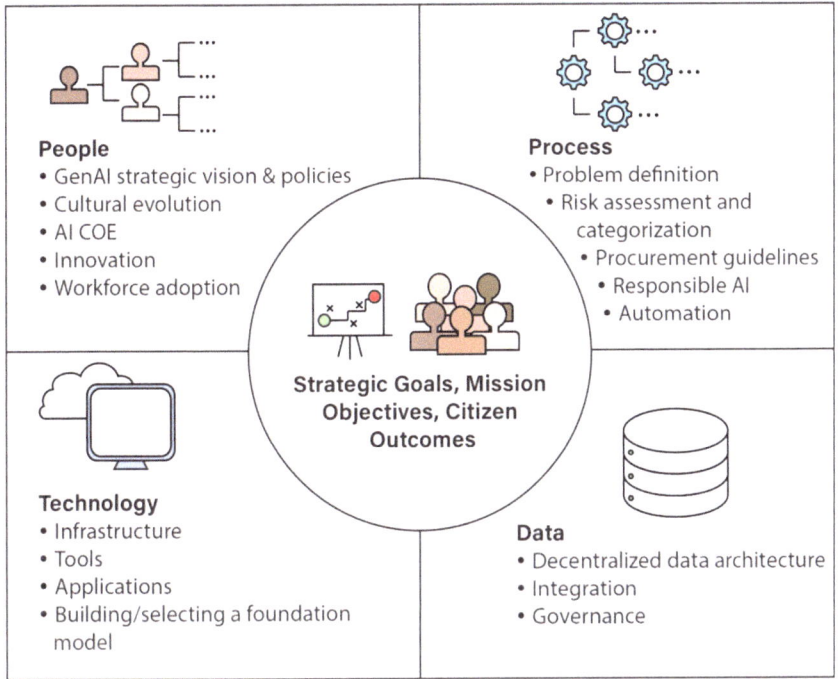

Figure 3-3. *Successful GenAI Adoption and Implementation: Four Perspectives*

As shown in Figure 3-3, the framework highlights essential considerations for each perspective. At the center of the framework are the strategic goals and objectives of the PSO that drive desired constituent outcomes. This illustrates that every perspective needs to align with the strategic goals and objectives of the PSO. Let's see how this framework helps in addressing the challenges we discussed.

3.4.1 Implementation from a People Perspective

What are the most difficult aspects to deal with in an organization? You probably guessed it: the people and the culture. These two are the most important determinants that drive the overall success of an organization. Therefore, our framework outlines several actions to make the workforce more comfortable reacting to change within the PSO in adopting GenAI:

- Establish the strategic vision and policies for GenAI

- Establish a path for gradual culture evolution toward GenAI

- Establish AI Centers of Excellence (ACOEs)

- Establish a culture of innovation

- Identify the right workforce

- Enable the workforce to adopt GenAI

Now, we can clearly see how having an overall strategy helps with GenAI adoption. The blueprint we mentioned earlier will enable PSOs to address several of these points. Let's go into more detail about each of these.

Establish the Strategic Vision and Policies for GenAI

Setting a strategic vision for GenAI and communicating that vision with key stakeholders in the organization will go a long way in setting up a successful GenAI journey. For example, let's say the PSO leadership settles on a vision that says "*Enhance constituent services where feasible using GenAI in a risk-averse manner.*" This vision provides the overarching scope within which GenAI must operate and sets the tone for implementation.

The vision also guides the development of policies that help govern the measured rollout of GenAI. For example, a policy aligning with the preceding strategic vision could be "*GenAI solutions shall be implemented with clear human oversight and control to ensure responsible decision-making."* Once the policies are laid out, each individual department or operating division can include goals and objectives that are outcome driven. These outcomes are related to constituent services and experience, staff efficiency and cost savings, and/or digital transformation. These goals align with the PSO's mission and have specific metrics or objective key results (OKRs) with a feedback loop to measure the success of GenAI initiatives.

Establish a Path for Gradual Culture Evolution Toward GenAI

How do you overcome the resistance to change when adopting new technologies, especially to mitigate the fear of job displacement or loss? The PSO leadership can create a culture of continuous evolution in which the agency workforce is comfortable with the transformational changes to improve the quality of constituent services. The leadership can achieve adoption of this culture – by clearly communicating the vision by leading with the value of this transformation to the mission. The leadership can help articulate the benefits of GenAI and that it is there to help improve productivity rather than replace the workforce.

Establish AI Centers of Excellence (ACOEs)

As mentioned earlier, buy-in from key stakeholders within the organization is essential to successfully adopt GenAI. One of the best ways to achieve this objective is to establish an ACOE. This entity should have stakeholders and senior leadership representation from various teams (business, technology, finance, legal, contracts, HR, and so on) to drive GenAI adoption and implementation within the agency. The ACOE sets the road map for GenAI within the agency, including prioritizing use cases, establishing governance, ensuring ethical AI implementation, and providing architecture best practices.

Establish a Culture of Innovation

Due to the risk-averse nature of the public sector, PSOs are typically behind the curve in adopting new technologies. Leaders can drive a culture of innovation by focusing on customer needs, challenges, and pain points and by using innovative technologies such as GenAI to solve these problems. Ultimately, innovation is about creating value for the constituents. One of the approaches is to experiment and fail fast using cloud-based services, which enables innovation without heavy upfront investments.

Identify the Right Workforce

GenAI adoption and implementation requires a variety of teams and personas with wide-ranging skills. Depending on your adoption and implementation patterns, including build or buy, customize, and fine-tune models, you need to consider the right team composition. The team includes personas to architect, design, develop, evaluate, and manage the GenAI applications. Table 3-1 lists the typical roles and responsibilities of these personas. (Note: We are not including resources required for training FMs.)

Table 3-1. *GenAI Team Roles and Responsibilities*

Role	Responsibilities
CIO (Chief Information Officer)	Strategic leadership: Identify public sector needs where GenAI can improve efficiency, personalize services, or automate tasks. Develop a strategy, road map for implementation and scaling GenAI within the organization.
IT Director	Technical implementation: Oversees platform build for deploying and managing GenAI models. Responsible for platform security, scalability, and accessibility.

(continued)

Table 3-1. (*continued*)

Role	Responsibilities
GenAI Architect	Design and optimize the overall architecture of the GenAI application, ensuring scalability, performance, and integration with existing systems.
Prompt Engineer	Design and optimize prompts that guide Foundation Models toward desired outputs. Build a library of predefined prompts, collaborate on user interfaces for effective prompt input, and work alongside data scientists and engineers to continuously refine prompts and ensure the GenAI application delivers the best possible results.
Data Scientist	Data and model management: Evaluate FM based on the use case and organization. Ensure data quality and security for fine-tuning and running GenAI models. Establish data governance policies and ensure compliance with regulations.
ML Engineer	Model development: Use-advanced machine learning techniques to fine-tune GenAI models. Collaborate with cross-functional teams to prepare data for model fine-tuning. Research and implement cutting-edge algorithms and architectures.
Policy Analyst/ Ethics Specialist	Governance and oversight: Mitigate potential risks associated with GenAI, such as bias, misinformation, and security vulnerabilities. Implement robust monitoring and control mechanisms. Address ethical considerations such as fairness, transparency, and accountability.
Program/ Project Manager	Collaboration and change management: Facilitate collaboration between IT teams and other departments for successful adoption of GenAI solutions. Develop and implement training programs. Address workforce concerns and facilitate adaptation to new workflows.

(*continued*)

Table 3-1. (*continued*)

Role	Responsibilities
Data Engineer	Responsible for data engineering tasks, including building and managing data pipelines for GenAI training and inference.
Full-Stack Developer	Responsible for developing the UI/UX, application orchestration, and database integration.
DevOps Engineer	Responsible for provisioning and managing deployments and operations of GenAI application, infrastructure, and related services.
Auditor	Independent assessment of GenAI models and audit of GenAI responses.
Security Specialist	Ensures that the GenAI applications adhere to the federal, state, and local security policies and other relevant regulations.

In later chapters, we will discuss how these roles are instrumental in GenAI solutions using specific use cases.

Enable the Workforce to Adopt GenAI

Workforce enablement, education, and transformation are critical for the success of GenAI initiatives. These include training the workforce on GenAI technologies and tools; educating them on the strengths and weaknesses of GenAI, risks, and mitigation strategies; and encouraging the workforce to innovate to fulfill the mission requirements. As part of this effort, PSOs can develop comprehensive training programs by partnering up with cloud and/or GenAI service providers.

3.4.2 Implementation from a Process Perspective

Developing standardized processes holds the key to solving many of the challenges and risks discussed earlier. Let's take a look at some important considerations.

Problem Definition and Benefit

The first and perhaps the most important consideration is to clearly identify the mission-related problem or opportunity. Scope out the opportunity in as much detail as possible and articulate realistic goals, benefits, and measures of success. The goals you define should tie back to the overall mission goals. At this stage, try to focus on the problem definition and opportunity to allow room for iterative development, where different solutions can be identified and continuously updated.

Risk Assessment and Categorization

Assessing risks and prioritizing the projects according to risk is another important step. Initially, you may want to find a project that has a lower risk or potential impact but at the same time can offer a glimpse into the benefits of GenAI. For example, let's say you are building a chatbot for improving productivity. An internal chatbot for employees has a lower potential risk than a chatbot to the external public. Once you've successfully built and deployed a project with a lower risk, you can use the lessons learned to take on a project at a greater scale and larger impact.

Procurement Guidelines

In the next section, we examine the different types of technology that a PSO may need to use for GenAI. It is important to ensure that prospective vendors provide the support necessary to address the risks of AI and GenAI. Vendors should clearly be able to articulate their commitment to the principles of responsible AI use, which is the next point we will discuss. Also, it is essential to implement a process for continued engagement with vendors for knowledge transfer. Ultimately, you want to upskill your staff to handle all the necessary functions to reduce your dependence on the vendor.

Responsible AI

Responsible AI is an umbrella term that refers to the practice of creating and using AI in ways that benefit society and minimize risk. Think of responsible AI as a set of principles that guide the development and deployment of AI systems, with the goal of building trust and ensuring these technologies are used in an ethical manner. Implementing responsible AI requires a proactive and intentional approach to ensure your systems are ethical, fair, and trustworthy. Responsible AI is an ongoing journey, not a one-time effort. Let's take a look at four of the main principles and practices to follow as part of this journey. We discuss examples and applications of these principles throughout the book.

1. **Ethical and regulatory compliance**: Due to risks associated with lack of transparency and explainability, the use of GenAI in public sector applications raises ethical questions, particularly in areas such as healthcare, law enforcement, and public communication. Decisions about the ethical boundaries of AI-generated content in these areas must be made thoughtfully. You may want to develop guidelines that restrict the implementation of GenAI in areas where the safety and rights of constituents come into play. Your guidelines should also align with any governmental strategies, policies, and improvement efforts. For example, consider the White House Executive Order on AI[6] released in October of 2023. This executive order requires state agencies to develop plans for the responsible use of all AI, including GenAI.

---

[6] www.whitehouse.gov/briefing-room/presidential-actions/2023/10/30/executive-order-on-the-safe-secure-and-trustworthy-development-and-use-of-artificial-intelligence/

2. **Error handling and transparency**: This aspect of
    responsible AI is essential for building public trust
    and ensuring accountability. Error handling and
    transparency should be prioritized throughout
    the GenAI life cycle. First and foremost, PSOs
    should establish robust monitoring and evaluation
    mechanisms to continuously monitor the
    performance of GenAI models by identifying errors,
    incorrect information, and other unintended
    consequences. Remember that an FM is purely
    generative and that the responses it provides can
    vary greatly. Establishing benchmark responses that
    set the standards to be used during evaluation is
    critical. In addition to this, when deploying a GenAI-
    based application, it is also critical to disclose to
    the users that an AI model is being used and there
    is a possibility of erroneous responses. You should
    also establish a mechanism for users to report such
    errors immediately so that corrections can be made
    quickly. Furthermore, you can implement reporting
    mechanisms to share findings and insights from
    monitoring and evaluation activities with the public.

3. **Bias and fairness detection**: Bias is the tendency
    of AI systems to reflect and amplify human biases,
    leading to unfair and discriminatory outcomes;
    however, don't confuse this type of bias with the
    terms bias in ML algorithms discussed in Chapter 1.
    AI system bias occurs in various ways, with the most
    prominent two being data bias and algorithmic
    bias. Data bias happens when the data used to train
    an FM is biased. For example, an FM trained on a

dataset of news articles that primarily feature men as CEOs may be more likely to identify a man as the CEO of a company compared to a woman. Algorithmic bias occurs when the algorithms used to build the FM are biased, even if the data used to train them is not. Because of this, when evaluating FMs for use, PSOs must examine the model cards, short documents that accompany FMs to provide a benchmarked evaluation in a variety of conditions. See Appendix D for examples of model cards.

4. **Content filtering and guardrails**: Earlier, we discussed the inherent risks with the input or prompts. These risks can be mitigated by implementing guardrails that handle prompts in your user interface. For example, your user interface can review prompts for malicious intent even before they are sent to the model. You can also create guardrails on the output. If, by any chance, someone is able to get the model to generate malicious content, then an output guardrail will detect this issue and invalidate the response. Your application should also encrypt all transmissions in transit between the user interface and the FM to prevent hacking. PSOs can also look into evaluating FMs on their ability to be manipulated using a process called red teaming,[7] simulating adversarial attacks to improve the robustness of FMs.

---

[7] https://hbr.org/2024/01/how-to-red-team-a-gen-ai-model

Process Automation

Automation is a powerful tool you can use in mitigating risks associated with GenAI, particularly in areas such as content safety and bias. You can train automated systems to flag potentially harmful content generated by AI, such as hate speech, misinformation, or deep fakes. This can be done by analyzing text, audio, or video for specific keywords, patterns, or inconsistencies. You can also use automation for testing and validating the output from FMs. Using automation frees up human reviewers to focus on complex or borderline cases that require more nuanced judgment and improves the overall efficiency of content moderation. However, when you automate processes, it is important to include adequate monitoring as well as human oversight.

3.4.3 Implementation from a Technology Perspective

In Chapter 1, we discussed how FMs are used in GenAI. For technology evaluation and selection, PSOs have a number of considerations. As a PSO, you need to determine how an FM is procured and deployed and then evaluate the infrastructure needed for model deployment. You may need to choose from and procure a large number of supporting tools, such as model evaluation tools and orchestrators (we'll talk more about them in the next chapter). Or, you could directly choose to implement an application that has already been developed by a provider. Let's look closely at the different considerations for each situation.

Building an FM or Selecting a Pre-trained FM

In most cases, you can leverage pre-trained FMs, such as Claude 3 from Anthropic, Titan from Amazon, GPT 3.5 and GPT 4.0 from OpenAI, and Google Gemini for building GenAI applications within a PSO. Building an FM from scratch is recommended only in a limited number of specific settings. Why? Because building an FM from the bottom up is a very resource-intensive option which takes a significant amount of time (months to years). It also requires large datasets, computational power, and expertise. Unless you need data that is highly specialized, domain specific, and not available on the Internet, there isn't a need to build your own FM.

Even if the data is somewhat specialized, you can use options, such as fine-tuning, to adapt the FM to a specific task or domain. This approach requires less data and resources. In fact, this is one of the reasons why GenAI is so powerful. You can use pre-trained FMs for your own tasks with great results.

How do you go about selecting a pre-trained FM, either for direct use or to use with fine-tuning? We provide a list of key tasks for doing so:

1. **Define clear objectives**: Clearly define the outcomes you want to achieve with the FM. Identify the use cases for specific problems in domains such as healthcare or education and tasks (summarization, question and answer, generation, chatbots, and so on) for which you will use the model. We will discuss use cases and tasks in great detail in Chapters 5–8.

2. **Evaluate model characteristics**: Examine the characteristics of the model using model cards (check Appendix D for a sample model card). A model card includes several details about the model. For example, it can include details such as the context window and the number of parameters. The number of parameters indicates the size of the model. The context window is the total number of tokens that an FM can process at one time, where a token is approximately equal to 75% of a word. This includes the input, output, and any history of previous conversations. Every model has a limit

on the context window. As the context window expands, the model can analyze more text, fostering a richer understanding and generating more cohesive responses. This mirrors how humans grasp context better through larger chunks of text, such as a paragraph, compared to isolated sentences. These details can play a role in how you develop your architecture, which we'll go into more detail in later chapters.

3. **Assess model performance**: Evaluate the model's performance using relevant benchmarks and metrics that are provided in model cards (see Appendix D for an example).

4. **Customization and flexibility**: Determine if and how the model can be customized or fine-tuned for your specific needs. Assess how easily the model can adapt to changing requirements or data.

5. **Scalability and efficiency**: Consider the computational resources needed to run the model, such as GPU and memory requirements. Ensure the model can scale according to your needs, especially if you are dealing with large datasets or require real-time processing.

6. **Ethical considerations and bias**: Evaluate the model for biases. Understand how it was trained to anticipate any skew in outputs. Consider the ethical implications of using the model, especially in sensitive applications.

7. **Community and support**: Having a strong community can be a valuable resource for support and troubleshooting. If the model comes from a vendor, assess the level of support and documentation available.

8. **Compliance and security**: Ensure the model complies with relevant regulations, especially for privacy and data protection. Understand any security risks associated with the model, particularly if using it in critical applications.

9. **Cost considerations**: Consider any costs associated with using the model, including licensing fees, computing resources, and maintenance. Weigh the costs against the expected benefits and ROI.

10. **Longevity and future-proofing**: Prefer models that are actively maintained and updated. Ensure that the model is likely to remain relevant and compatible with future technologies and standards.

By thoroughly evaluating these factors, you can make a well-informed decision in selecting an FM that not only meets your current requirements but is also sustainable and effective for future needs.

Procuring a Pre-trained FM

PSOs have the following options for procuring a pre-trained FM:

- **Open source models**: There are a number of models such as Llama, Jurassic-1 Jumbo, and CLIP, many of which are available on Hugging Face, that offer low cost and flexibility but might require more technical expertise for fine-tuning and deployment.

- **Commercial licenses**: Companies including Anthropic and OpenAI, to name a few, offer pre-trained models with various capabilities and license options, providing support and features such as fine-tuning APIs.

- **Cloud AI platforms**: Major cloud providers, including Amazon Web Services, Google Cloud, and Microsoft Azure, offer subscription-based access to pre-trained models and managed services for training, fine-tuning, and deployment. These include proprietary models such as Amazon Titan and Google Gemini as well as commercial models from Anthropic and OpenAI.

Navigating the FM landscape can be overwhelming with so many options! We have provided some resources in Appendix A that can help you to make a choice. However, we need to keep in mind that this landscape is rapidly evolving, as models and model versions are being released all the time.

So how should PSOs go about choosing between the various types of models? The choice comes down to each specific use case and the resources and culture of the PSO. We provide some of the factors that PSOs can consider as follows:

- **Transparency and control**: Open source models such as Llama offer complete access to the code, allowing for customization and independent audits. This can be crucial for PSOs with concerns about privacy, security, and potential biases in the model. Another way to increase transparency is through watermarking.[8] This is a process of embedding a recognizable, unique signal into the output of the FM, such as text or an

---

[8] www.brookings.edu/articles/detecting-ai-fingerprints-a-guide-to-watermarking-and-beyond/

image, to identify that content as AI generated. That signal, known as a watermark, can then be detected by algorithms designed to scan for it.

- **Performance and capabilities**: Third-party models from Anthropic and OpenAI may have superior performance due to factors such as access to proprietary data and training methods. Consider whether the slight performance edge outweighs the transparency benefits of open source models for your specific use case.

- **Cost and resources**: Open source models eliminate licensing fees, but require in-house expertise for deployment, maintenance, and potentially fine-tuning. Third-party models often come with support packages and managed services, but at a cost. Evaluate your internal resources and budget constraints.

- **Data security and privacy**: PSOs have stringent data privacy regulations. Assess how both open source and third-party models handle data. Open access in open source models might require additional security measures to ensure sensitive data isn't exposed. Third-party models may offer stronger data security guarantees depending on their specific policies.

- **Focus and alignment**: Consider the specific focus of the FM. Some open source models, such as Llama, might be more general-purpose, while third-party models might be tailored toward specific domains relevant to the public sector (e.g., healthcare, education, legal).

Again, it is very important for a PSO to assess in-house expertise for deploying and potentially fine-tuning open source models. Even with in-house expertise, it might be difficult for organizations to keep up with the quickly changing landscape. That is one of the biggest advantages of using cloud-based platforms such as Amazon, Azure, and Google. You can get access to a wide range of models that you can pilot with minimum investment to assess suitability and performance in your specific context.

Infrastructure

As discussed earlier, ultimately, most PSOs would be looking to either directly deploy a pre-trained FM or fine-tune a pre-trained FM. Deploying these models requires performant, cost-effective infrastructure that is purpose-built for ML. Most costs are associated with running the models and doing inference. GenAI applications can constantly generate predictions, known as inferences, in real time and require very low-latency and high-throughput networking. Amazon Alexa is a great example, with millions of requests coming in every minute, which accounts for 40%[9] of all computing costs.

Although PSOs can procure this infrastructure on-premises, it would require significant hardware investments and technical expertise for managing the infrastructure. For that reason, we recommend deploying models on cloud platforms that provide the necessary scalability, flexibility, cost optimization, and high throughput required. The other advantage with a cloud-based model is that the PSO can easily use the latest infrastructure. This is a considerable advantage given that GenAI is an emerging technology, and things are changing very quickly.

---

[9] https://aws.amazon.com/blogs/machine-learning/announcing-new-tools-for-building-with-generative-ai-on-aws/

Tools for Building an Application

In Chapter 4, we will discuss the different parts of a GenAI application. In order to build these applications, you may need additional tools such as vector databases and orchestrators. There are many choices for these tools, from open source to commercial. Many cloud providers also make these capabilities available as part of their existing database offerings. The choice ultimately comes down to integration and compatibility with the FM provider. We provide a list of FM providers in Appendix A.

Out-of-the-Box Applications

There are many providers offering ready-made GenAI capabilities out of the box. This is a great option to consider for specific use cases as it eliminates all the grunt work necessary to develop and maintain an application. For example, GitHub Copilot is an AI assistant integrated into GitHub that helps developers write code, suggests completions, and refactors existing code, and it has the potential to accelerate software development. In this case, the choice comes down to your specific business needs and the solution's compatibility with your infrastructure and the regulations to which you must adhere. We provide a list of these GenAI applications in Appendix B.

3.4.4 Implementation from a Data Perspective

We've looked at best practices from three other perspectives (people, process, and technology). However, we haven't covered one of the most important and difficult challenges in a PSO: implementation from a data perspective.

The concept of a "data-driven organization" is here to stay because so many companies embrace data in their day-to-day decision-making. As the saying goes, "Data is the new oil," and it is critical within the public sector especially for decision-making. The McKinsey Global Institute estimates[10] that data and analytics could create approximately $1.2 trillion a year in value across the public and social sectors.

Public sector leaders have to rely on data-driven decision-making for a variety of initiatives. Just to name a few, these initiatives include

1. Analysis of the quality of constituent services

2. Federal compliance and reporting

3. Policy formulation and analysis

4. Budget and resource justification and allocation

5. Risk management

6. Program or IT operations including controlling fraud, waste, and abuse

7. Empower constituents with access to their own data

Data is the most important organizational asset for PSOs; after all, almost anything else, including people, equipment, processes, and technology, can be replaced or replicated. However, you can only imagine what it means to a PSO to lose data that it has collected over decades. Data also becomes the foundation for innovation and transformation within any PSO.

How is this relevant to GenAI? Well, we've all heard the familiar phrase "garbage in, garbage out," and that applies to GenAI as well. If you want to leverage GenAI for use cases within your PSO such as chatbots, search, and summarization, you need to manage the large amounts of data and that is a major challenge. Data is also critical if you want or need to fine-tune an FM or mitigate bias within GenAI. Therefore, your data has to be up to date, complete, accurate, discoverable, and available when needed. For many PSOs, this is a complex and challenging task. Let's briefly consider why:

---

[10] www.mckinsey.com/industries/public-sector/our-insights/
accelerating-data-and-analytics-maturity-in-the-us-public-sector

- **Data volumes and types**: The US public sector is potentially the largest producer of data,[11] with information collected by thousands of federal, state, and local agencies across the areas such as public health, census, revenue or taxes, and public safety, surveillance, and intelligence, just to name a few. For example, data. gov has over 250,000 datasets on various topics. PSOs collect, store, and process constituent data of all different types in large quantities – hundreds of terabytes or even petabytes. For example, we discussed earlier that, within the United States, there are hundreds of millions of beneficiaries across multiple programs, such as Social Security, Medicare, Medicaid, and so on. This results in hundreds of terabytes of different types of data collected every year across these programs in the form of benefit applications, notices, appeals, etc., for just one of these benefit programs alone.

- **Data quality**: Data quality is a major challenge particularly given the volume and disparate types of data available in both structured and unstructured (audio, videos, images) formats. Poor quality data leads to poor models – data quality issues include duplicates, missing values, incorrect formats, and so on.

- **Data security and privacy**: Given that PSOs collect constituent data, setting policies, such as data access, ownership, and accountability, in compliance with the regulations is critical. Agencies face many challenges with data privacy and security.

---

[11] https://digital.gov/2018/03/14/data-briefing-value-federal-government-data/

- **Other challenges**: Data silos, lack of data sharing, and data-driven insights for decision-making with inflexible data warehouses and traditional one-size-fits-all relational databases.

While discussing a holistic data strategy is beyond the scope of this book, we'll narrow down the discussion in the context of GenAI and address the question "What do PSOs need from a data perspective to succeed with GenAI initiatives?" There are three main areas that PSOs should focus on:

- **Decentralized data architecture**: Due to strict regulatory compliance, data security, and privacy concerns, PSOs often have data silos, each managed by a data team with a lot of data duplication. This creates inflexibility in your mission agility. You can resolve these issues by adopting a decentralized data architecture. In this pattern, data is treated as a product with each business unit having full autonomy and ownership of their data domain. The business units can share the data across the organization using a decentralized governance framework, such as a data mesh. Also, in this context, due to regulations, if data sharing is not possible, you can use the concept of *federated learning*[12] which enables organizations to build models without exchanging data. The discussion of federated learning is outside the scope of this book.

- **Data integration**: A decentralized data architecture implies a lot of data sharing and integration. This integration involves combining data from disparate

---

[12] https://research.google/blog/federated-learning-collaborative-machine-learning-without-centralized-training-data/

sources and formats for analytics, reporting, and
program compliance. This process also involves
options to extract, transform, and load the data. For
example, a consumer of data should be able to access
source data in place using queries. These technologies
typically use data source connectors that can help
harmonize data across different data sources.

- **Data governance**: A decentralized data architecture
  requires data governance to enable business producers
  to share their data with consumers. The data should
  be up to date, complete, accurate, discoverable,
  and available when needed. So how should you go
  about this?

Building a robust data governance framework is the first step. This
involves establishing clear policies and standards that dictate how data is
sourced, used, and protected throughout the GenAI life cycle. Every step
needs to be governed by ethical principles and responsible data practices.
Technology plays a crucial role in streamlining data governance, with
specialized platforms helping monitor data quality, detect anomalies, and
enforce privacy compliance. Integrating robust security measures such
as encryption and access control further strengthens the defenses against
unauthorized access or misuse of data and generated outputs.

However, technology alone is not enough. Fostering a culture of data
governance within an organization is equally vital. This means educating
everyone involved in GenAI projects – from developers to executives –
about data privacy, security, and the ethical considerations that guide
data usage. Open communication channels, where concerns can be freely
raised and addressed, further uphold this culture of shared responsibility.

Remember, data governance isn't a static set of rules but a dynamic
journey of continuous improvements. Regularly auditing and assessing
your data governance practices ensures that they stay relevant and

effective in the ever-evolving world of GenAI. By embracing data governance wholeheartedly, PSOs can unlock the vast potential of GenAI while remaining steadfastly committed to responsible, ethical, and sustainable development.

3.5 Conclusion

Adopting GenAI in the public sector presents a unique set of challenges and considerations across people, processes, technology, and data. However, by taking a strategic and responsible approach, PSOs can harness the immense potential of GenAI to drive better constituent outcomes and improve workforce productivity.

The key is to establish a clear vision aligned with the agency's mission, build a culture of innovation, and develop robust governance mechanisms that prioritize ethical AI practices. Investing in the right talent, infrastructure, and data strategies is crucial for successful implementation. By following the methodology and framework outlined in this chapter, organizations can mitigate risks and navigate the challenges faced while ensuring compliance with relevant regulations.

Ultimately, the adoption of GenAI is an ongoing journey that requires continuous monitoring, fine-tuning, and a commitment to responsible AI principles. As the technology rapidly evolves, PSOs must remain agile and adaptable, continuously reassessing their strategies and practices to stay ahead of the curve.

In the next chapter, we'll look more closely at the technical aspects of building GenAI applications. We'll explore key concepts such as prompts, prompt engineering, and best practices for developing robust and effective GenAI solutions. Understanding these fundamentals will allow you to explore GenAI-driven tasks and use cases that unlock multiple benefits for PSOs.

CHAPTER 4

Building a Generative AI Application

In Chapter 1, we briefly introduced Foundation Models (FMs). In this chapter, we will explore the key aspects and components of Generative AI (GenAI)–based applications. Some of the concepts we will discuss in this chapter are

- The anatomy of a prompt, which acts as the interface for communicating with Generative AI models.

- The art of prompt engineering, revealing techniques to craft clear and effective prompts that guide FMs toward desired outputs.

- The concept of agents, showcasing how they can facilitate complex reasoning and multistep problem-solving.

- Various techniques to handle domain-specific data such as Retrieval-Augmented Generation (RAG) and model fine-tuning.

- A generic application architecture for a GenAI application.

By the end of this chapter, you will understand all the core concepts necessary for building a GenAI application in the public sector.

© Sanjeev Pulapaka, Srinath Godavarthi and Dr. Sherry Ding 2024
S. Pulapaka et al., *Empowering the Public Sector with Generative AI*,
https://doi.org/10.1007/979-8-8688-0473-1_4

4.1 Anatomy of a Prompt

Figure 4-1 depicts the basic interaction of a user with an FM.

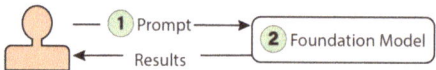

Figure 4-1. *Basic Request to a Foundation Model*

In the diagram, the user makes a request to an FM, and the FM provides a response. This request is called the *prompt.* But similar to a conversation with a human, the clearer you are, the better the results, making the format of the prompt crucial. We illustrate the general anatomy of a prompt in Figure 4-2.

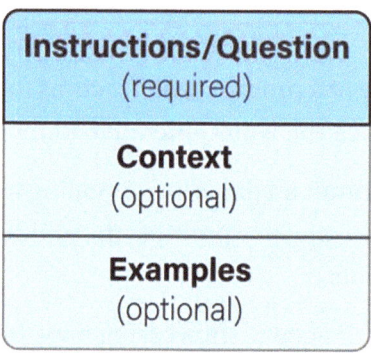

Figure 4-2. *The Anatomy of a Prompt*

You can break down a prompt into three broad sections as shown in Figure 4-2:

1. **Instructions or question**: This is text that tells the model what you want to do.

2. **Context**: The *context* is additional information
   that helps guide the model toward a more
   comprehensive understanding of the task and its
   nuances. While not always required, *context* can
   significantly enhance the accuracy and relevance
   of the model's output. Typically, it would include
   relevant background information or domain-
   specific knowledge, previous conversations or
   interactions with the model, or specific instructions
   or preferences regarding the desired response.

3. **Examples**: Provide examples of similar tasks to
   get the model to answer a question. Again, this is
   not required but can be very helpful for complex
   scenarios.

The total length of the prompt, including instructions, context, and
examples, can vary from model to model and even within a model. Some
models may require additional mandatory components or words. For
example, Anthropic Claude requires you to open the prompt by using the
keyword Human.

Note   It is important to keep in mind that the format of a prompt is
not fixed.

The preceding format is simply a guideline. For example, you can add
other things such as the role you want the FM to assume and the expected
output format. Let's say you are using the FM to write a contract. You
can ask the FM to assume the role of a contract writer for developing a
contract. If you are trying to use the FM to explain a complex concept, you
can ask the FM to assume the role of a teacher to explain a concept to a
student.

Given the wide range of possibilities in constructing a prompt and its critical nature in determining the output from a model, an entirely new field has emerged called prompt engineering.

4.2 Prompt Engineering

We introduced prompts in the previous section. Prompts are critical to control the output from the FM. An improperly constructed prompt will not give you the best results. The importance of having an optimally constructed prompt cannot be understated. The field of prompt engineering is dedicated to establishing best practices and refining the art of constructing these prompts.

There are several types of common prompting techniques used in prompt engineering. Being an active field of research, newer techniques and frameworks are continuously evolving.

Let's have a look at a few techniques used in text-to-text generation models or Large Language Models (LLMs). In each example included as follows, we provide the description of the technique, the structure of the prompt using the example in Figure 4-2, and the output generated by the LLM.

4.2.1 Zero-Shot Learning

This technique allows the model to complete a task without any specific examples, relying solely on its understanding of language.

Prompt:

```
Instruction: Write a social media post summarizing a new grant
program for small businesses.
Context: NA
Examples: NA
```

Output:

Attention small business owners! We're excited to announce a new grant program to help businesses recover from recent natural disasters. Funding is available for repairs, equipment replacement, and more. Find out if you're eligible and apply today!

Zero-shot learning is especially useful when you want the model to be creative and come up with new ideas.

4.2.2 Few-Shot Learning

This technique provides a small number of examples to the LLM, allowing it to learn from those examples and generalize to new instances. If you provide one example, it is called one-shot, two examples two-shot, and so on. The example in the following prompt is therefore two-shot.

Prompt:

Instruction: Create a concise safety reminder about public transportation.
Context: Focus on reminding riders to be aware of their surroundings.
Examples:
Input: Always hold on to railings while on the bus.
Output: Stay safe on the bus! Remember to hold on to railings for a smooth ride.
Input: Look out for closing doors when exiting the train.
Output: Exiting the train? Mind the gap and wait for doors to fully open before stepping out.

Output:

Be Aware, Ride Safe! When using public transportation, stay alert and keep your belongings close. Let others know if you need assistance.

By giving specific examples, you are helping the model provide more accurate answers and reducing the scope for *hallucinations*, which are incorrect or misleading results. These results can be caused by a variety of reasons, including insufficient training data, incorrect assumptions made by the model, or biases in the data used to train the model.

4.2.3 Chain-of-Thought Prompting

This technique encourages the model to break down a complex problem into smaller steps and provide its thought process by giving it a few examples. Let's see how this works for a student trying to solve a math problem.

Prompt:

```
Instruction: Solve the following math word problem by breaking
it down into steps and showing your work.
Context: The agency case worker has a 40-hour work week.
Each case takes 4 hours to complete. How many cases can they
complete in 40 hours?
```

Examples:

```
Question:"An employee works for 30 hours, and needs 5 hours to
complete a task. How many tasks can they complete in 30 hours?"
```

Answer:

```
1. We are given that the employee has 30 hours
2. Each task required 5 hours
3. To find the number of tasks they can complete, we divide the
total number of hours by the number of hours needed for each
task: 30 ÷ 5 = 6.
Therefore, the employee can complete 6 tasks in 30 hours
```

Output:

To solve this problem, we need to follow these steps:
 1. The agency case worker has a 40-hour work week.
 2. Each case takes 4 hours to complete.
 3. To find the number of cases the worker can complete in 40 hours, we divide the total number of work hours by the hours needed for each case: 40 ÷ 4 = 10.
Therefore, the agency worker can complete 40 cases in 10 hours.

As you can see from the output, using this method provides clear guidance to the LLM and makes it easier to understand and verify the reasoning of the output.

4.2.4 Prompt Chaining

This technique involves breaking down a complex task into smaller subtasks and using the output of one subtask as the input for the next subtask, effectively chaining multiple prompts together.

Prompt 1:

Instruction: Identify the key points from a constituent's complaint about a malfunctioning traffic light.
Context: The constituent submitted a complaint through an online form, reporting a broken traffic light at a specific intersection.
Output 1: Malfunctioning traffic light reported at [intersection name].

Prompt 2 (using Prompt 1 output):

Instruction: Draft a short message acknowledging the complaint and outlining the next steps.

Context: The traffic light is located on a major road with high traffic volume.

Output 2:

Thank you for reporting the malfunctioning traffic light at [intersection name] on a major road. We understand the urgency and a crew will be dispatched to assess the situation as soon as possible. Expect updates within the next 2 hours.

4.2.5 Prompt Templates

Prompt templates are a mechanism that allows you to standardize the prompt language while changing the inputs. This allows you to reduce the overhead of creating new prompts for each task especially if the task is repetitive only with a few changes that include different variables.

Let's say you want the LLM to generate biographies of a number of different scientists. You can create the following prompt template:

Instruction: Create the biography for the given scientist.
Context:
        Name: {Scientist}
        Lived: {Dates}
        Known for: {Major Achievements}
        Interesting Fact: {Key Information}]

You provide the model with the template and the name of the scientist, let's say Albert Einstein. The model recognizes the template as a structure to follow. It fills in the blanks with information about Albert Einstein as follows:

Name: Albert Einstein
Lived: 1879 - 1955
Known for: Theory of Relativity
Interesting Fact: Won Nobel Prize for photoelectric effect

The model creates the following output:

```
Albert Einstein (1879 - 1955) was a German-born theoretical
physicist widely acknowledged as one of the greatest and most
influential physicists of all time. He is best known for
developing the theory of relativity, one of the two pillars
of modern physics (alongside quantum mechanics). His work is
also known for its influence on the philosophy of science.
An interesting fact about Einstein is that although he won
the Nobel Prize in Physics in 1921, it was not for his theory
of relativity, but for his explanation of the photoelectric
effect, a crucial step in the development of quantum theory.
```

Notice that the details from the template are used to create a coherent and informative narrative. When you feed in the names of each of the scientists, the model will create biographies that contain this information, ensuring some kind of consistency between outputs. Another benefit is that you don't need to create each prompt multiple times.

4.2.6 ReAct (Reasoning + Acting)

The ReAct framework, described in ReAct: Synergizing Reasoning and Acting in Language Models[1], extends the chain-of-thought approach we discussed earlier. The core idea is to simulate human-like problem-solving by interspersing the reasoning steps with actions that gather additional information. Let's see how this works.

Similar to the chain-of-thought approach, you provide an LLM with a prompt that includes several examples. However, the format of the example is provided as follows.

---

[1] https://arxiv.org/abs/2210.03629

Prompt:

```
Instruction: The instruction or question for the LLM
Context: relevant context if needed
Examples: Several examples of the task presented in the
following manner:
```

Example1:

```
Instruction or Question: The instruction or question
for the LLM
Action: A specific action the LLM can take to gather
information (e.g., "Search the web for...").
Observation: The expected outcome or information gained from
the action.
Thought Process: Steps the LLM should consider while taking the
action (e.g., "Consider the context of the question...").
```

Example2:

```
Instruction or Question: The instruction or question
for the LLM
Action: A specific action the LLM can take to gather
information (e.g., "Search the web for...").
Observation: The expected outcome or information gained from
the action.
Thought Process: Steps the LLM should consider while taking the
action (e.g., "Consider the context of the question...").
```

When this prompt is passed to the LLM, the LLM uses examples as the guide to accomplish the required task. This is especially useful when there are a number of different options that the LLM can consider to answer a question and to determine which option to choose. We explore this technique in the next section ("Agents").

4.2.7 Agents

Fundamentally, *agents* are independent systems that implement dynamic reasoning using LLMs. Agents use prompt engineering techniques such as ReAct, which we covered earlier, to guide the LLM through the reasoning process. Agents also have various tools at their disposal to obtain observations that are needed for the LLM.

Let's say you are part of a public sector organization (PSO) responsible for improving public health. You want to investigate areas with limited access to healthy food options in a particular city or country. You have several data sources within the organization that contain relevant data, such as food store inventory, pricing data, maps, locations, and accessibility information. Using agents, you can interact with an LLM so that you can ask questions and draw useful insights. Agents have tools, such as content management connectors, database connectors, or Application Programming Interfaces (APIs), at their disposal. They would then use the LLM to figure out the type of content needed for a question, use the tool to get the content, and then use the LLM to put together the answer.

The following relatively simple example illustrates the use of agents based on the ReAct prompt engineering technique.

You ask an LLM this question:

```
Which constituents have purchased more than $1000 worth of
products in the last month?
```

The information about customers and products is in a database. To get this answer, the LLM needs to know that it has to retrieve the data from a database. It also needs to know which database, which tables, and specifically which columns to pull the required information. Because of the complex reasoning required and the fact that a database connection is needed, you choose to answer this question using an agent.

The agent can connect to the database and run queries. Also, since we are using the ReAct framework, the agent sends a prompt to the LLM with a series of examples. The prompt contains the question and a series of examples using the Thought-Action-Observations pattern. This is used to get answers from data in a database, as we discuss the ReAct prompt engineering section.

When the prompt is received, the LLM attempts to solve the problem using the same Thought-Action-Observations pattern as in the example. The agent keeps examining the output from the LLM, and if an action, such as connecting to the database, is needed, it performs that action and provides the result back as an observation to the LLM. The interaction looks similar to the text outlined in Table 4-1.

Table 4-1. *LLM Thought-Action-Observations Pattern*

LLM	Agent
Thought: I need to identify customers with > $1000 in products last month.	
Action: Identify the customers.	
Observation: Data resides in a database; I cannot access it directly.	
Thought: I need to connect to the database to get the answer.	
Action: Connect to the database.	Parses LLM output and determines that a connection to the database is needed.

(continued)

Table 4-1. (*continued*)

LLM	Agent
	Connects to the database API.
	Confirms connection to the LLM.
Observation: I am connected to the database.	
Thought: I need the purchase date and spending per customer.	
Action: Get the purchase data and spending per customer.	Parses LLM output and determines a query needs to be executed.
	Executes query based on LLM request.
	Provides purchase data and spending per customer to LLM.
Observation: I have a list of customers and their spending.	
Thought: Calculate total spending and identify those exceeding $1000.	
Action: Calculate total spending and identify those exceeding $1000.	
Thought: I have the list of customers exceeding the $1000 threshold.	
Action: Finished – Here is the list of customers exceeding the $1000 threshold.	Parses LLM output and determines final answer has been reached.
	Presents the list to the user as the final answer.

This example showcases the potential of this approach to tackle complex public sector problems that require intricate reasoning and multifaceted data analysis. In more complex problems, the LLM would have multiple options to perform different types of actions, and the agent would have multiple tools available at its disposal. The LLM would iteratively try different options until it finds the right answer.

Currently, there is active ongoing research in this field, and the results show a lot of promise. The ability to construct a system that can reason and act independently is a major step forward in the pursuit of Artificial General Intelligence.

4.3 Best Practices for Constructing Prompts

The preceding methods are just a few examples of prompt engineering techniques for LLMs; there are many more variations and combinations that can be explored to improve the performance and capabilities of language models. Prompt engineering for text-to-image models is different from text-to-text and uses slightly different techniques, including negative prompts. We'll explore this in more detail in Chapter 5.

The goal of prompt engineering is always the same: guide the LLM to yield the desired results. The effectiveness of each technique greatly depends on the specific model and the task. You can also combine multiple techniques to achieve optimal results.

We discussed various techniques for creating prompts earlier, but in order to be effective, you also need to know how to construct each prompt, regardless of the technique used. Let's look at a few guidelines with examples.

4.3.1 Clarity and Specificity

Be precise and clearly specify what you need from the model. Vague prompts may lead to vague or irrelevant responses. For example:

Vague prompt:

```
Analyze trends in constituent satisfaction.
```

Clear prompt:

```
Analyze trends in constituent satisfaction ratings for public
transportation in the past five years, and identify areas for
improvement.
```

While more information can be helpful, overly long prompts may confuse the model. Strive for a balance when providing context and details. This helps the FM understand the nuances of the request.

4.3.2 Structured and Logical Format

Present your request in a logical, structured manner. Bullet points or numbered lists can help in complex queries. If your prompt involves multiple steps or concepts, present them in a logical sequence. For example:

Complex prompt:

```
Research and write a report on the economic impact of the new
recycling program.
```

Structured prompt:

```
Gather data on recycling rates before and after the program
implementation.
```

Analyze the data to identify any correlation between recycling rates and economic indicators.

Write a report summarizing the findings and potential economic impact.

4.3.3 Understanding FM Capabilities and Limitations

Understand the FM's capabilities and limitations, including model parameters such as temperature and context window, the ability to work with multiple languages, or the ability to work with both text as well as images. Here are a couple of examples:

Prompt:

Generate a social media post in Spanish promoting the new childcare program.

This considers the model's ability to handle multiple languages.

Prompt:

Design a new website for the city government using this large dataset.

This prompt might exceed the model's capability for complex tasks.

4.3.4 Iterative Approach

Begin with a broader query and then refine it based on the FM's responses to zero in on your exact need. Use the responses to tweak your prompts, learning how changes in your query affect the output. For example:

Initial prompt:

Provide a summary of recent traffic accidents in the city.

Refined prompt (based on initial response):

Based on the previous summary, identify high-accident zones and suggest potential safety measures.

4.3.5 Use of Examples

Use few-shot prompting with specific examples for more complex ideas or desired outcomes. For example:

Prompt:

Write a press release announcing the launch of a new grant program for small businesses. Here's an example of a press release from last year... Adapt it to the new program details.

4.3.6 Avoiding Biases and Assumptions

Use unbiased, neutral language to avoid leading the FM toward potentially biased or incorrect outputs. Ensure your prompt does not inadvertently contain assumptions that could skew the response. For example:

Biased prompt:

Identify the best neighborhoods for young families to live in.

Note that this prompt assumes a specific definition of "best."

Neutral prompt:

Analyze factors that influence families' choices of neighborhoods, considering aspects such as education, housing costs, and amenities.

4.3.7 Creative Use of Prompts

You can instruct the FM to assume a certain role (e.g., a tutor, a specific expert) to shape its responses. Frame your query in the context of a hypothetical scenario for more focused responses. Here is an example:

Prompt:

```
Assume the role of a financial advisor and write a budget plan
for a single parent on a low income.
```

4.3.8 Ethical Considerations

Craft prompts that encourage ethical and responsible use of AI. Avoid prompts that could lead to harmful, biased, or unethical outputs.

Unethical prompt:

```
Write a social media post encouraging people to use a
new facial recognition technology for crime prevention,
highlighting its effectiveness without mentioning potential
privacy concerns.
```

Ethical prompt:

```
Write a balanced social media post introducing the new facial
recognition technology for crime prevention. Briefly explain
its benefits and potential drawbacks, such as privacy concerns.
Encourage constituents to learn more and participate in a
public forum to discuss its implementation.
```

By mastering the art of constructing prompts, you can significantly enhance the effectiveness and accuracy of FM model interactions. Experimentation is the key! As LLMs continue to evolve, so too will the art of prompt engineering. Try different methods and evaluate responses to see which method works best for which model and task.

4.4 Model Parameters and Configurations

Beyond crafting the prompt itself, various model parameters also heavily influence the output generated by a Large Language Model (LLM). These parameters can differ across different LLMs. We briefly explored some of them in Chapter 1, but let's dive deeper with public sector–specific examples.

4.4.1 Temperature: The Creativity Knob

Imagine temperature as a dial controlling the creativity of the LLM's output. A lower temperature setting prioritizes factual accuracy, making it ideal for tasks including

> **Public health communication**: When generating public advisories about a new disease outbreak, a low temperature (around 0.0) ensures clear and accurate information delivery.

> **Legal document summarization**: Summarizing complex legal documents for constituents requires factual accuracy. A low temperature setting minimizes creative embellishments.

Conversely, a higher temperature (around 1.0) injects more randomness, leading to creative and surprising outputs. This might be useful for brainstorming public engagement initiatives: for example, a high temperature setting can help generate diverse ideas for engaging constituents in community discussions.

4.4.2 Top-k Sampling: Focusing the Output

Top-k sampling restricts the LLM's vocabulary choices during text generation. Consider only the top k most probable words at each step. Here's how k values can impact the output:

- **Lower k value (e.g., k=10)**: When generating grant application summaries for policymakers, a lower k value ensures a focused and concise output, highlighting key points.

- **Higher k value (e.g., k=50)**: When generating different policy options for a complex social issue, a higher k value allows the LLM to explore a wider range of possibilities, providing policymakers with diverse perspectives.

4.4.3 Top-p Sampling: Balancing Control and Exploration

Similar to top-k sampling, top-p sampling focuses on a specific probability range for selecting the next word. This technique considers only words with a cumulative probability falling within a certain threshold (top_p). Choosing between top-k and top-p depends on your specific needs:

- **Lower top_p value (around 0.6)**: Imagine generating constituent feedback summaries for a transportation project. A lower top_p value prioritizes the most common themes in the feedback while allowing for some exploration of less frequent, but potentially valuable, insights.

- **Higher top_p value (around 0.9)**: When writing a press release about a new environmental regulation, a higher top_p value allows for some creative phrasing while still prioritizing clear and accurate communication of the regulation's purpose.

Remember, these configurations significantly impact the output. Consulting the LLM provider's documentation is crucial for optimal parameter settings.

4.5 Model Evaluation

For PSOs looking to adopt Generative AI in a responsible manner, it is critical to carefully evaluate the outputs of FMs. Given that this is an evolving field, there is no gold standard for evaluation techniques. However, PSOs can review several types of metrics to evaluate FMs, depending on the specific task or application. Here are some commonly used metrics:

- **Perplexity**: Perplexity is a measure of how well an LLM predicts a given sequence of words. Lower perplexity values indicate better performance.

- **Toxicity**: Measures presence of harmful content in the output.

- **Bilingual Evaluation Understudy (BLEU)**: BLEU is a widely used metric for evaluating machine translation systems. It measures the similarity between the model's output and reference translations. Higher BLEU scores generally indicate better translation quality.

- **Recall-Oriented Understudy for Gisting Evaluation (ROUGE)**: ROUGE is a set of metrics used for evaluating text summarization systems. It measures the overlap between the generated summary and reference summaries.

- **Accuracy**: Accuracy measures the proportion of correctly classified instances out of the total number of instances. It is a straightforward metric but can be misleading in imbalanced datasets or tasks with multiple classes.

It is important to note that the choice of evaluation metric depends on the specific task and the characteristics of the data. In many cases, multiple metrics are used to gain a comprehensive understanding of the model's performance from different perspectives.

While automatic metrics are useful for quick evaluation, human evaluation is probably still the best method for assessing the quality of FMs. Human evaluators can provide more nuanced feedback and better capture aspects such as fluency, coherence, and overall quality. In addition to human evaluators specifically assigned to review models, feedback from end users of the system can also prove to be extremely useful. In this case, you can choose between different types of rating methods such as star rating, thumbs up/down, and numeric rating on the GenAI application user interface.

4.6 Techniques to Handle Domain-Specific Data

At this point in the book, you should understand that FMs are pre-trained with large sets of generally available data. However, what if you want an FM to answer questions on specific data within your organization? In that case, there are three options:

- You can find a way to pass your data as context to the model. One method to do this on a consistent basis is called Retrieval-Augmented Generation (RAG).

- You can further train the model for your specific needs and data. This is called fine-tuning.

- You can build your own custom FM.

Building your own custom FM is a very expensive, time-consuming, and complex option. Therefore, in most cases, we recommend using fine-tuning and RAG. Each of them has their own benefits and disadvantages. Let's first take a look at RAG.

4.6.1 Retrieval-Augmented Generation (RAG)

As we mentioned earlier, RAG is a technique used for passing domain data as context to an FM in a consistent and repeatable manner. It applies more to LLMs and Multimodal LLMs (MLLMs) as opposed to image generation models. Figure 4-3 illustrates how RAG works.

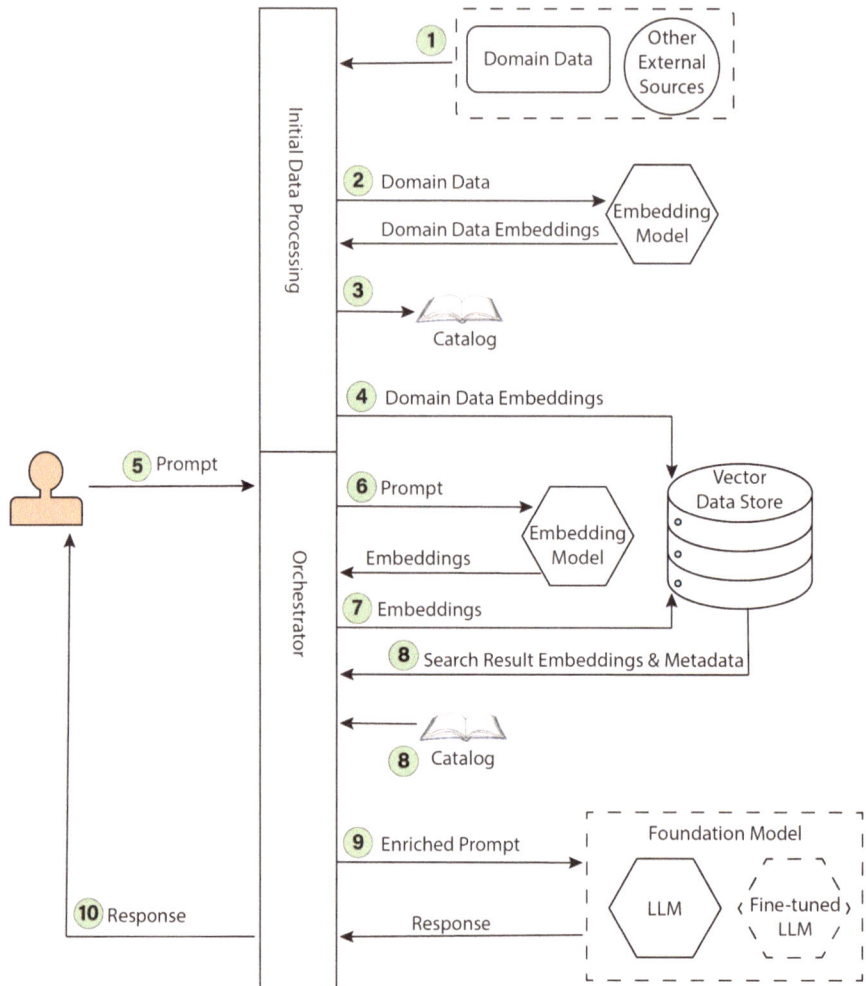

Figure 4-3. *Retrieval-Augmented Generation (RAG)*

Figure 4-3 shows each step in the RAG process from start to finish. The overall process has two phases.

Initial data processing phase: In this phase, you convert the domain data into a format that is suitable for optimized retrieval of information. Whenever the source data changes, you need to repeat the ingestion phase to ensure that all data is updated. Let's take a look at each step in the process:

1. Domain data is retrieved from miscellaneous data sources such as document repositories, databases, or APIs.

2. The data is broken into smaller manageable chunks and sent to an embedding language model. The model converts the chunks of data into embeddings. *Embeddings* are numerical representations of the data within a vector space.

3. A catalog of the embeddings and corresponding chunks of text is maintained.

4. The embeddings are stored in a database, called a *vector database*, specifically designed to store such data. There are many vector databases, including one named Pinecone, available in the market. Cloud-based database services such as RDS from AWS and Azure Cosmos also offer this capability.

Execution phase using an orchestrator: This phase is the interaction between a user and an LLM. When a user creates a request, the request is matched to the domain data, and the most relevant pieces of information are retrieved. This data is then passed as context to the LLM along with the request. The LLM then provides a response to the request but most importantly uses the context as the basis of information.

In most cases, you would use an orchestrator to execute this workflow. An orchestrator is software that manages complex workflows and interacts with an FM using a prompt and an optional context. Let's walk through the steps of this phase:

5. The orchestrator receives the prompt from the user. Let's call this the initial prompt.

6. The orchestrator sends the initial prompt to the embedding model to convert the text into embeddings. Note that for MLLMs, images accompanying the text will also need to be embedded.

7. The embeddings of the query are compared with the embeddings in the vector database using a similarity search algorithm.

8. The search results are converted back into text by comparing the retrieved embeddings with the chunks of data in the catalog.

9. The orchestrator appends the initial prompt with the search results. Let's call this the "enriched prompt" to differentiate from the initial prompt. The enriched prompt is sent to the FM. The FM can be a pre-trained model or a fine-tuned one.

10. The response generated by the FM based on the enriched prompt is sent back to the user.

This technique opens up a vast number of possibilities and use cases within the corporate world.

A major benefit with RAG is the ability to increase accuracy and reduce hallucinations. However, in order to do so, your data should be accurate in the first place! And your prompts need to be engineered using some of the techniques discussed earlier.

RAG is a complex technique and involves a number of steps as seen before. Therefore, it is important to evaluate the quality of RAG responses to ensure that the model is performing as required. Next, we discuss some methods you can use to evaluate the responses.

4.6.2 Evaluating RAG Responses

The quality of RAG responses is crucial for its use in GenAI-related applications. Faithfulness, relevancy, and accuracy are three measures that can be used for this. Let's break down each term:

- **Faithfulness (also known as groundedness)**: This refers to how closely the generated answer adheres to the factual information present in the retrieved context. A faithful answer accurately reflects the retrieved passages without fabricating or distorting the information. High faithfulness ensures that the answer is grounded in truth and reliable.

- **Relevancy**: This describes how pertinent the generated answer is to the specific question asked. A relevant answer directly addresses the query and provides information that is helpful for understanding it. While an answer can be faithful to the retrieved context, it might not be relevant if it doesn't answer the specific question effectively.

- **Accuracy**: This encompasses both faithfulness and relevancy, evaluating whether the generated answer is both correct and relevant to the question. An accurate answer provides true information that answers the query directly. It is the ideal outcome, combining factual grounding with the desired focus.

It is important to note that these qualities can interplay. High faithfulness doesn't always guarantee high relevancy. An answer might accurately cite the retrieved context but not directly address the specific query, making it irrelevant. High relevance might involve slightly compromising faithfulness. To answer the question directly, the model might need to slightly adapt or summarize the retrieved information, impacting its strict faithfulness but still providing a useful response.

Understanding these dynamics helps us determine RAG's effectiveness in different scenarios. When accuracy is important, as it is in factual queries, both faithfulness and relevancy become crucial. However, for open-ended questions, relevancy might take precedence over strict faithfulness, as long as the answer provides insightful and helpful information.

In addition to purely metric-driven measures, it can help to add human evaluation into the mix. This would mean creating benchmarked gold standard responses for standard questions against which a human can compare and provide ratings.

By analyzing these qualities through appropriate metrics, you can evaluate the strengths and weaknesses of RAG models. This ultimately leads to improved effectiveness and more informative and impactful responses.

4.6.3 FM Fine-Tuning

Fine-tuning an FM is the process of adjusting a pre-trained model to perform better on specific tasks or with specific types of data. This process can significantly enhance the model's relevance and accuracy for your particular use case.

Fine-tuning can be divided into two broad areas:

1. **Domain adaptation**: The objective here is to improve the FM's ability to understand domain-specific data, such as industry-related terms, technical terms, or other specialized languages.

2. **Instruction based**: The objective in this case is to improve the FM's ability to complete a specific task. Instruction-based fine-tuning uses labeled examples. The labeled examples are formatted as prompt and response pairs and phrased as instructions.

In both these cases, you can adjust the weights of an FM in such a way to provide better performance for your specific problems/objectives. Traditional methods for doing so include methods we have discussed before, such as supervised, unsupervised, or reinforcement learning with human feedback (RLHF). However, many organizations need to own the process of fine-tuning a model using their own data. Because FMs typically have billions of parameters (weights), adjusting these weights in the traditional method used for other models typically isn't feasible for end customers or organizations. Instead, a method called Parameter-Efficient Fine-Tuning (PEFT) has been gaining prominence. PEFT adjusts only a small subset of parameters, thereby significantly reducing the computational cost and memory requirements for fine-tuning. Popular PEFT techniques include Low-Rank Adapters (LoRA) and Quantized Low-Rank Adapters (QLoRA).

Fine-tuning does come with some additional work and overhead on your part; so, before embarking on fine-tuning, it is important to first thoroughly evaluate the performance of the base pre-trained model to determine if it can handle the task or recognize your data. We recommend using the RAG approach first using your corporate data. If you determine that the pre-trained model with RAG does not meet your requirements, then you may need to use fine-tuning. Here are some best practices to consider for fine-tuning:

1. **Understand the pre-trained model**: Familiarize yourself with the capabilities and limitations of the base model. Understand what data the model was originally trained on to gauge its inherent biases and strengths.

2. **Define clear objectives**: Have a clear understanding of the outcome you want to achieve through fine-tuning. This will guide your data selection and fine-tuning strategy. For example, let's say a PSO wants to create a chatbot system to answer frequently asked questions (FAQs) about a new social welfare program. The PSO can fine-tune a pre-trained LLM using a dataset specifically tailored to the new program. This dataset would include official program documents, FAQs and informational materials, and transcripts from past call center interactions about similar programs.

3. **Curate a high-quality dataset**: Ensure that the fine-tuning dataset closely aligns with your target domain or task. Use a diverse and sufficiently large dataset to capture the nuances of your specific use case. In general, diversity is more important than

the size of the data. This means the data should comprehensively represent the different variations across the use case. Make sure to clean and preprocess the data to eliminate noise and irrelevant information.

4. **Start with a small subset**: Begin with a small subset of your data. This allows for quicker iterations and helps in understanding how the model responds to fine-tuning. Gradually increase the dataset size based on initial results and identified needs.

5. **Regular evaluation**: Use a separate validation set to regularly evaluate the model's performance during fine-tuning. Monitor for signs of overfitting or underfitting and adjust your approach accordingly.

6. **Incorporate domain expertise**: Where possible, involve domain experts in reviewing and annotating the fine-tuning dataset. Use expert feedback to refine the model iteratively.

7. **Ethical and bias considerations**: Actively look for and mitigate biases in both the training data and the model's outputs. Consider the ethical implications of the model in your specific domain and strive for responsible AI use.

8. **Testing with real-world scenarios**: Test the fine-tuned model with real-world scenarios to ensure it performs well in practical applications. Gather user feedback to further refine the model.

Fine-tuning an FM is a delicate process that balances enhancing the model's performance on specific tasks with maintaining the model's general capabilities. By following these best practices, you can effectively tailor the model to your specific needs while ensuring its reliability and ethical use.

4.6.4 RAG vs. Fine-Tuning

Now that we have seen both RAG and fine-tuning, you may be wondering how you could distinguish and choose between the two. The answer is it really depends on the use case and the particular scenario. There are a number of different aspects you can keep in mind when comparing and evaluating the two techniques for your use case. We summarize those techniques in Table 4-2 and follow up with more details on each aspect.

Table 4-2. *Evaluating Fine-Tuning and RAG*

Aspect	Fine-Tuning	RAG
Focus	Improves LLM performance on specific tasks within existing data.	Improves LLM performance on diverse tasks through access to external knowledge.
Data Usage	Primarily relies on additional task-specific data.	Utilizes both pre-trained LLM data and external knowledge sources.
Resource Intensity	Resource-intensive during training due to additional data processing.	May require more resources for initial setup and retrieval depending on knowledge base size.
Applications	Ideal for tasks requiring adaptation to specific datasets (e.g., sentiment analysis for product reviews).	Ideal for tasks requiring access to specific, up-to-date, or domain-specific knowledge (e.g., question answering, generating legal documents).
Flexibility	Requires retraining the LLM for different tasks or domains.	More adaptable to new tasks by adjusting the knowledge base and prompt engineering.
Explainability	May not be clear where the information in the response originates from.	Can include citations or references to sources, increasing explainability and user trust.

Focus

- **Fine-tuning**: Imagine training a government analyst to specialize in a specific policy area. Fine-tuning focuses an LLM's capabilities on a particular task within its existing knowledge base. For example, an LLM analyzing public sentiment could be fine-tuned on datasets of constituent tweets related to a specific government program.

- **RAG**: Think of RAG as equipping the analyst with a research library. RAG broadens an LLM's knowledge by allowing it to access and integrate information from external sources such as government documents and databases.

Data Usage

- **Fine-tuning**: This approach relies heavily on curated datasets specific to the target task. Let's say a department wants an LLM to identify fraudulent unemployment claims. They would fine-tune the LLM using historical claim data with confirmed fraudulent cases.

- **RAG**: RAG leverages the LLM's pre-trained knowledge along with information retrieved from the external knowledge base. An LLM could be linked to a database of past legal rulings to assist with drafting regulations, drawing on both its understanding of language and relevant legal precedents.

Resource Intensity

- **Fine-tuning**: Training can be resource-intensive. The LLM needs to process and be "fed" the additional data. Imagine the analyst needing to read a vast amount of policy documents to become a specialist.

- **RAG**: Setting up and maintaining a large knowledge base can require significant resources. Additionally, retrieving information during LLM generation takes processing power. Think about the time it takes a researcher to find relevant resources in a library.

Applications

- **Fine-tuning**: This method excels when you need the LLM to become an expert for a specific dataset and task. For example, an LLM could be fine-tuned on historical grant proposals to improve its ability to categorize new proposals for funding.

- **RAG**: RAG is ideal when the LLM needs access to specific, up-to-date, or domain-specific knowledge. An LLM could be linked to a real-time database of weather patterns to generate tailored advisories during natural disasters.

Flexibility

- **Fine-tuning**: This often requires retraining the LLM for different tasks or domains, which can be time-consuming. Imagine the analyst needing to be retrained for a new policy area.

- **RAG**: RAG offers more flexibility. You can adjust the knowledge base and how you prompt the LLM to adapt it to different situations. It is similar to having a constantly updated research library at your disposal.

Explainability

- **Fine-tuning**: It might be unclear where the information in the LLM's response comes from, making it difficult to assess its accuracy.

- **RAG**: RAG can include citations or references to the knowledge sources used, allowing users to understand the basis of the response and fostering trust.

Both fine-tuning and RAG are valuable tools for enhancing LLM performance in public sector applications. The best choice depends on your project's specific needs and resource constraints.

4.7 GenAI Application Architecture

So far, we have discussed some important concepts such as prompts, RAG, and agents that are needed to handle some of the complexities and nuances of a GenAI application. Figure 4-4 depicts the application architecture of such an application.

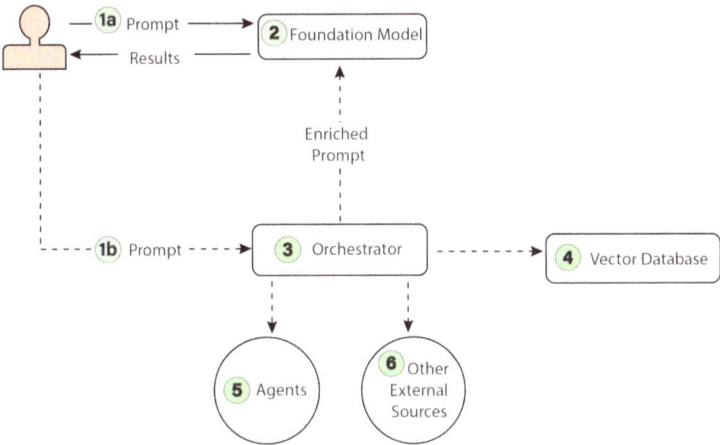

Figure 4-4. *GenAI Application Architecture*

Let's discuss each step of the architecture. Some steps are optional depending on your business needs and the model you're considering. For example, text-to-image generation models may not need components 3, 4, and 5.

1. **Prompt**: As discussed earlier, the prompt contains instructions as well as additional information to help the model provide a relevant answer. The prompt can be sent (1a) directly to the FM or (1b) to the orchestrator.

2. **FM**: The FM that performs your generation task. It receives the prompt directly from the user or from an orchestrator and generates a response, passed either directly back to the user or to the orchestrator. It is important to evaluate and select the right model for your use case. The FM itself could be pre-trained or fine-tuned depending on your domain and the

characteristics of the task at hand. See Chapter 3, section 3.4.3, for additional guidance on selecting a model.

3. **Orchestrator**: We discussed the orchestrator briefly in our section on RAG. In the case of RAG, the orchestrator sends a prompt enriched with the relevant context to the FM. Orchestrators may also add additional statements to the prompt sent by the user. This may include guardrails or statements that ensure that the FM does not provide answers to questions deemed as harmful by the organization. Orchestrators also help with other complex activities such as handling multiple FMs, handling agents, or making an external API call.

4. **Vector database**: As a PSO, you typically store a lot of information in documents. Examples are contracts, procedures, resumes, policy information, and so on. As discussed earlier in the RAG section, you can store these documents in the vector database for retrieval.

5. **Agents**: Agents, as explained in the previous section, will help you work with an LLM on complex reasoning problems.

6. **Other external sources**: These are additional sources outside the organization such as websites and APIs that the orchestrator can call depending on the use case.

This architecture lays the foundation for building GenAI applications for a variety of tasks including document and image generation, chatbots, and summarization and reporting. In later chapters, we explore variations in this architecture specific to these tasks and use cases.

4.8 Conclusion

In this chapter, we explored the core components and underlying concepts for building GenAI applications. We covered the anatomy of prompts, the art of prompt engineering, model parameters, RAG, fine-tuning techniques, and the concept of agents. By understanding and effectively utilizing these elements, you can unlock the power of GenAI models in your PSO to tackle a wide range of challenges.

However, remember that GenAI is not a one-size-fits-all solution; each task or domain may require specific considerations and tailored approaches. In the next chapter, we examine one of the most promising applications of GenAI in the public sector: content generation. We explore how GenAI can be leveraged to create high-quality, personalized, and engaging content across various mediums, from text to images and multimedia. By understanding the nuances of content generation, you can enhance your PSO's communication, education, and outreach efforts, ultimately better serving your constituents and advancing your mission.

CHAPTER 5

Content Generation

In Chapter 2, we discussed briefly about different applications that GenAI can be used for, including content generation, summarization, chatbots, search, and reporting. In the next four chapters, we will go into greater detail on each of these applications using concepts described in Chapter 4 such as prompt engineering and Retrieval-Augmented Generation. In this chapter, we will describe how GenAI applications can help with content generation tasks.

Public sector organizations (PSOs) face many challenges with content generation. These challenges range from document to image to code generation.

Let's take the example of Pablo, a lead contract administrator in a PSO. Pablo finds that his team spends a lot of time creating contract-related artifacts. The information his team needs is spread all across the organization. Also, his team has a number of regulatory requirements to deal with including the Federal Acquisition Regulations (FAR).[1] With tight budgets, he has constraints on the number of people he can hire.

Sofia, the communications specialist, needs the PSO content to adhere to strict legal and regulatory guidelines, ensuring accuracy and compliance across all formats. She understands the importance of making the content

---

[1] `www.acquisition.gov/browse/index/far`

accessible and inclusive for all constituents, some of which was mandated by law. This means considering different languages, cultural contexts, and accessibility needs.

Maria, the Chief Information Officer (CIO), faces an uphill battle. Her team, including Mary, the IT director, constantly struggles with heavy workloads and limited resources. The programmers in her team complain that a number of their tasks are repetitive and tedious, and this has resulted in increased staff turnover.

Jie, a training instructor, finds it challenging to engage students without creating compelling visuals. However, creating these visuals takes up a lot of her time.

The preceding examples represent a subset of the many challenges PSOs face in real life. GenAI offers promising solutions to address these types of challenges in three broad areas: document generation, image generation, and code generation. In this chapter, we will discuss these in detail along with high-level architectures and sample use cases.

5.1 General Areas of GenAI Content Generation

5.1.1 Document Generation

Document generation with GenAI offers great potential to PSOs. By automating routine tasks such as contract generation, document summarization, and content translation, GenAI can free up valuable staff time and resources. Additionally, GenAI ensures consistent and accurate content creation by adhering to regulations and checking for factual information. Furthermore, GenAI promotes accessibility by generating content in various formats and simplifying complex language, fostering inclusivity for all constituents. So, Pablo, the lead contracting officer, and

Sofia, the communications specialist, don't have to spend time on these types of activities and can focus on other larger mission areas.

From generating contracts to public appeals and notices, the applications of this capability are quite diverse. There are a number of tools in the marketplace that offer these capabilities out of the box; we list them in Appendix D. While existing tools offer convenience, they often impose limitations on customization and control. Hence, you may need to consider a custom implementation using FMs. This helps tailor the retrieval and generation processes to your specific needs; it also helps manage how information is sourced, filtered, and integrated into generated documents.

5.1.2 Image Generation

Image generation with GenAI unleashes some exciting possibilities for the public sector, from creating synthetic images for medical research to transforming complex data into impactful visuals for education and awareness campaigns and fostering design concepts for planning and engagement. Creating high-quality images through traditional means can be time-consuming and resource-intensive, and maintaining consistent quality while producing large volumes can be difficult.

GenAI overcomes these limitations, offering speed and efficiency in creating images. GenAI models can be customized to generate specific styles and variations, expanding your creative possibilities and offering a level of customization not easily achievable with traditional methods. This is a significant benefit for training professionals such as Jie, from our hypothetical PSO.

5.1.3 Code Generation

PSOs grapple with resource limitations and the need for efficient, secure, and accessible software solutions. Code generation tools powered by FMs offer a compelling solution. These tools automate repetitive tasks, such as code generation, freeing up developers for complex problem-solving. They also ensure consistent, high-quality code by adhering to coding standards and mitigate human error. For the CIO and the IT director, this translates to timely delivery of services to constituents through faster development cycles, reduced costs, and better customer experience through improved code quality and reduced defects.

Now that we've briefly discussed each of the three broad areas where GenAI can assist with content generation, let's dive deeper into each of these areas. Let's begin by examining document generation.

5.2 Document Generation

As discussed earlier, document generation with GenAI offers a lot of potential. There are out-of-the-box solutions as well as custom-built solutions for document generation. However, given the specific nature of documents within the public sector, PSOs may need a custom solution for document generation in many cases. Let's first look at a high-level architecture of a customer document generation solution. We then discuss two use cases that apply this architecture.

5.2.1 High-Level Architecture

Figure 5-1 illustrates the high-level architecture of a document generation application.

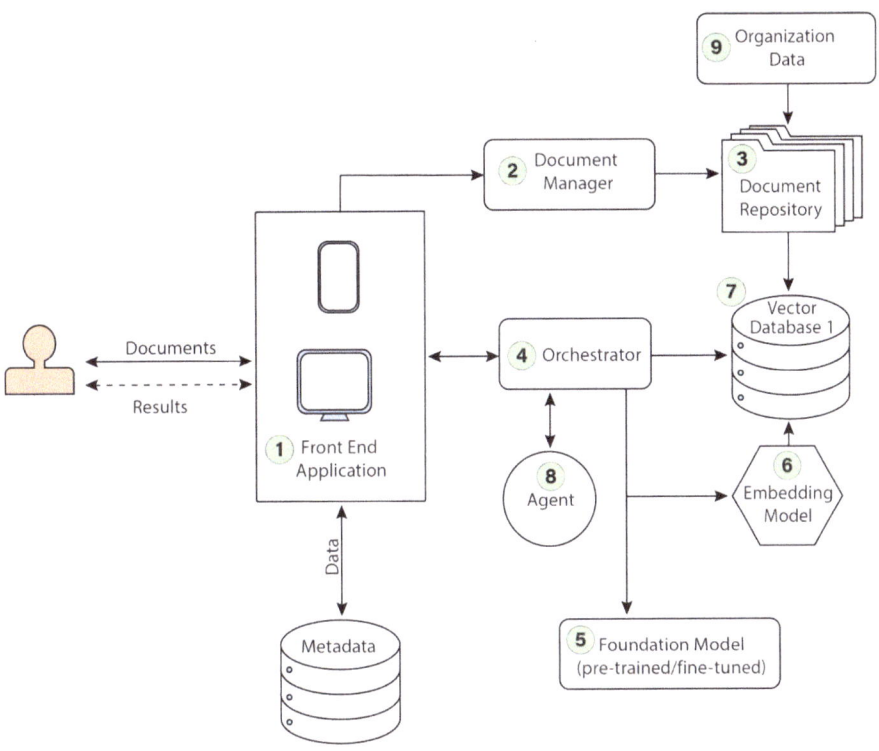

Figure 5-1. *A Solution to Generate Documents Using GenAI*

This architecture is an extension of the application architecture discussed in Chapter 4 and illustrated in Figure 4-4. At a high level, this architecture allows users to generate different types of documents such as contracts, project charter, program management reports, architecture/ design, and test plans, to name a few. Users specify a prompt on a front-end application. This prompt is sent to an FM which generates either a full document or sections of a document. Users can then review, edit, store, and retrieve these documents or sections.

However, you can see that the architecture contains multiple components. Let's understand the role of each component in more detail:

1. **Front-end application**: The user interacts with a front-end application that presents a user interface (UI). The user could be a person seeking to generate a document or a reviewer reviewing existing documents. Once the initial document or section is generated by the FM, the user should be able to edit the document. You can have manual reviews as well as automated reviews by sending the generated content back to another FM to review the accuracy of the content. As such, the user interface should have the ability to flag sections of content that are not accurate or satisfactory. These flags have to be stored in a metadata repository so that you can use the information for fine-tuning the prompts. The user should have the ability to upload a document or template that is used as a reference for generating the content. The documents uploaded could be in various formats such as a PDF, a spreadsheet, or a document file, and so on.

2. **Document manager**: The front end interfaces with a document manager that is responsible for formatting and storing documents/content generated from the FM. For example, when new content is generated by the FM, the user might want to see that content as a PDF, a spreadsheet, or a document file. In this case, the document manager does the job of formatting the content. The document manager also manages documents and templates uploaded by the user using the front-end application.

3. **Repository**: The repository stores and organizes the content. This could be new content generated by the FM, documents or templates uploaded by the user as reference, as well as additional reference documents obtained from a PSO's internal systems. These systems contain information needed as context when the user makes a request to the FM. The information could be contained in unstructured data such as documents in various formats or structured data such as databases.

4. **Orchestrator**: The front-end application invokes an orchestrator (described in the previous chapter) that does the heavy lifting by interacting with the FM for content generation. You can create content for each section of the document, the entire document, or multiple times for each section. Note that different types of prompts are needed for the different types of documents created. The orchestrator implements logic to pass in the appropriate prompt for different types of documents. We recommend creating prompt templates, which we discussed in Chapter 4, for different types of documents so that you can reuse the prompts. You can also create templates for generating each section within a document.

5. **FM**: This can be an FM of your choice, and you can use either the base model or a fine-tuned model. As an example, to generate contract-related documents, fine-tuning can be done using procurement-related documents or nomenclature specific to procurements.

6. **Embedding model**: As described in Chapter 4, this is an optional component that is needed for passing relevant context to the FM using RAG, which we described in detail in Chapter 4. The embedding model converts document text from the document repository to vectors.

7. **Vector database**: We need a vector database to store and retrieve embeddings (vectors) generated by the embedding model.

8. **Agent**: As described in Chapter 4, an agent enables you to integrate with other tools to access organization data. For example, if data is in a relational database, then the agent guides the FM to create Structured Query Language (SQL) to extract the required information. An agent can also be used to directly interface with APIs external to the organization.

Now that we have reviewed the main components, let's discuss some of the key considerations that a PSO needs to keep in mind when using this architecture.

5.2.2 Key Considerations

Front-End Application UI

The UI is one of the most important considerations for any application, and GenAI applications are no exception. As with any other application, design the UI based on established techniques that keep customer experience as the top focus. This includes best practices, such as ensuring that the UI adapts seamlessly to different screen sizes and devices and making the UI accessible for users with disabilities, adhering to Web Content Accessibility Guidelines (WCAG).

Handling Large and Complex Document Generation

While there are some use cases where content generation is fairly brief, there are several others that involve large and complex documents, such as contracts. In these cases, it is critical to break down the content into smaller sections and use prompt templates designed for each section. When creating content, you should prioritize creating crucial sections first. This ensures continuity across sections because you can pass initially generated content as relevant context to subsequent sections.

Multiple Models

PSOs are not limited to using just one FM for content generation. In fact, using multiple FMs opens up several possibilities. For example, you can use one FM to generate content and another to check the quality of that content. Or you can use two FMs to produce the same content and then use metrics to evaluate the better one. As the landscape for FMs grows, this will invariably become a necessity.

Human Review

In Chapter 1, we discussed how FMs work and the chance that the content generated is not accurate all the time. Therefore, it is essential to include human review in the content generation process. PSOs need to determine the type and frequency of quality checks needed as well as the points where checks are needed. This is based on factors such as the business use case and the criticality of the application. For example, the content distributed to constituents may need additional reviews, or procurement/legal-related documents may need multiple reviews for each section. Keep in mind though that human review isn't just about fixing errors; it is also about optimization. Think of the GenAI tool as something that can get you 70–80% of draft content, but a human is required to get it to 100%.

Data Quality

The architecture shows different types of data sources, including organization data repositories, data uploaded by a user, or external APIs or websites. The quality of this data is important, so establish processes to ensure the same. We'll discuss these GenAI implementation considerations in detail in Chapter 9.

Cost

Finally, it is essential to understand the implications of sending data as context to an FM. Sending large amounts of data can increase cost and latency. We discuss costs and cost optimization in detail in Chapter 9.

Now that we have an understanding of the architecture, we can discuss different use cases for content generation across the public sector. In each use case, we discuss the different roles/personas involved in implementing the solution. We outlined these roles in Chapter 3. (Note that the roles are meant to be a high-level guide; the actual team composition will be much more comprehensive.)

5.2.3 Document Generation Use Case 1: Contracts and Procurement

Problem

The primary mode through which PSOs acquire products and services is through contracting and procurement processes. As an example, in fiscal year 2022, the US federal government obligated \$694B[2] through contracts. Anyone who is involved in the procurement process at a PSO understands how time-consuming and resource-intensive the process is. According

---

[2] https://gaoinnovations.gov/Federal_Government_Contracting/

to a Gartner survey,[3] with an average of 22 months, the public sector has the longest buying cycles for technology purchases compared to other industries.

In our example, let's take Pablo, the lead contracting officer at a hypothetical PSO. Pablo faces significant challenges in gathering the necessary information to create well-structured, comprehensive, and compliant materials. These include Request for Proposals/Information (RFPs, RFIs), Statements of Work (SOWs), contracts/contract modifications, and purchase orders. The information required to create these contracts can be scattered across multiple departments and data sources. Pablo needs to spend a considerable amount of time and resources researching and collecting this information.

Another problem Pablo faces is compliance with the many process requirements and guidelines within each organization as well as across the government. For contracts in the United States, he needs to comply with the Federal Acquisition Regulation (FAR)[4] which governs procurement for federal agencies. However, in the European Union, his team needs to comply with the Public Procurement Directive,[5] a common framework for public procurement across member states. Last but not least, he has to make sure that international and cross-country procurements comply with regional trade agreements such as the North American Free Trade Agreement (NAFTA) and the European Union's Internal Market rules on public procurement.

---

[3] www.gartner.com/en/newsroom/press-releases/2022-09-06-gartner-survey-finds-government-tech-purchase-decisio

[4] www.gsa.gov/policy-regulations/regulations/federal-acquisition-regulation-far

[5] https://eur-lex.europa.eu/legal-content/EN/TXT/PDF/?uri=CELEX:32014L0024

Solution

The CIO determines that a GenAI-based solution can help Pablo overcome the preceding challenges and formulates a cross-functional team. This team implements the high-level architecture described in Figure 5-1.

Developers design an intuitive front end that Pablo's team can use to generate contract and procurement documents. The front end will also enable Pablo to upload supporting documents if required. They ensure that the document manager interfaces with the PSO's existing contract management systems. They also implement a workflow which the CEO can use to review and sign-off on contracts.

Data scientists evaluate the various FMs and select the most appropriate FM for this use case using guidelines provided in Chapter 3.

Data engineers develop processes to upload contracts and procurement-related templates, previous examples of contracts, and policies to the repository. Information from these documents is embedded into a vector database and passed as context to the FM using the RAG approach.

Prompt engineers work on optimizing the prompts necessary to generate the contracts required by Pablo. Since the system uses the RAG approach, the prompts would include necessary context from the repository or from documents uploaded directly to the front end. For example, if an RFI for an IT system is needed, the prompt could get relevant data for the IT system from the repository. As another example, if the RFI has to comply with certain rules and regulations, then the relevant rules would be obtained from the repository and passed as context to the FM.

Developers and prompt engineers work interactively with Pablo to test the results. They iteratively develop and review different types of prompts for different types of contracts.

Outcome and Benefits

The solution helps improve productivity and efficiency of the entire contract team at the PSO. This solution also enables Pablo's team to generate draft contract documentation that complies with regulations.

Overall, using GenAI, Pablo's team has been able to cut down the procurement times drastically, acquire products and services in a timely manner, and help deliver services to constituents faster with lower costs.

Note that there are additional benefits – the GenAI application developed can be customized for different departments, agencies, or even jurisdictions based on the relevant regulations and laws that apply. This ensures procurement documents adhere to the appropriate rules and requirements. The application uses content from previous RFP responses and contracts if required to meet compliance needs. The application also highlights where requirements may be unclear or inconsistent with past procurements. This allows Pablo's team to improve and refine specific clauses that traditionally vendors have struggled with.

Additional Considerations

Let's look at some additional factors that the PSO needs to consider in this use case:

- **Maintain audit trails for regulatory purposes**: To ensure transparency, accountability, and compliance, Pablo needs to maintain detailed audit trails of the document generation process, including the prompts used, the data sources accessed, and the versions of the AI model employed.

- **Continuously improve prompts based on user feedback**: The effectiveness of the document generation process heavily relies on the quality of the prompts provided. Developers need to establish a feedback loop in the front end to allow Pablo to provide input on the generated documents. For example, Pablo can indicate the quality of the contracts using a thumbs-up/down or a rating scale. Prompt engineers can then use this feedback loop to refine and evolve prompts.

- **Scale to support high volume of documents**: Pablo's team has to often deal with a high volume of contracts and procurement documents. So, the architect needs to ensure that the solution can scale efficiently, handling large volumes of requests without compromising performance or accuracy. We will discuss these implementation considerations in more detail in Chapter 9.

- **Integrate with e-signature and publication workflows**: Once the documents are generated, they typically need to go through additional processes, such as electronic signatures and publication. Developers should integrate the front-end application with existing e-signature and publication workflows to further optimize Pablo's time. This integration can involve APIs, data exchange protocols, and other mechanisms to ensure a smooth transition between the different stages of the document life cycle.

- **Fine-tuning**: Data scientists and ML engineers may need to evaluate fine-tuning of the FM if prompt engineers do not get the desired results.

- **Ensure data privacy and security**: Security specialists should ensure that the solution implemented by developers ensures robust data privacy and security measures. This may involve techniques such as data encryption, access controls, and secure data transmission protocols. Developers can also use other FMs to identify and redact sensitive procurement information when sharing documents with vendors. We will discuss more about security and privacy in Chapter 9.

- **Provide training and support**: Introducing a new document generation application powered by GenAI may require you to make significant changes in workflows and processes for procurement teams. To ensure a smooth transition and maximize the benefits of the system, it is essential for the CIO to ensure that her team provides comprehensive training and support to Pablo. This can include user guides, tutorials, and dedicated support channels to address any questions or issues that arise during the adoption phase.

This use case demonstrates how a PSO can effectively leverage the power of GenAI for document generation in the context of contracts and procurements while ensuring compliance, scalability, and seamless integration with existing processes.

5.2.4 Document Generation Use Case 2: Public Communication, Alerts, and Notices

Problem

PSOs often send out similar letters, emails to tens of thousands to millions of people. They also need to create alerts and notices for important events. As an example, post pandemic, state Medicaid agencies had to

137

communicate with over 90 million people to keep them enrolled[6] in the program. This communication involved renewal letters, text messages, emails, etc.

GenAI can personalize these communications with relevant details, improving engagement and making information more relevant to individual constituents.

Let's imagine a hypothetical PSO that represents a country's state department. Xiulan, the PSO's administrator, needs to send out responses to constituents who periodically send emails with questions about visas, passports, and many other state-related requests.

Sofia, the communications specialist, needs to work with embassies all over the world to create alerts and notices of events. She has to ensure that the messaging is clear and concise, and it needs to comply with various regulations. However, since each situation is unique, developing customized content is time- and resource-intensive. Information is often spread across departments, making it difficult to quickly compile accurate details. Review cycles can also delay publishing critical information.

Solution

The CIO decides to use GenAI to help address Xiulan's and Sofia's challenges. Her team implements the high-level architecture described in Figure 5-1.

The developers in the team design an intuitive front end that Sofia can use to generate alerts for critical events. They ensure that the document manager interfaces with the PSO's existing alert management systems. However, this time, they also develop a process that can schedule automatic generation of responses to questions from constituents. They do this by integrating the email system with the orchestrator. They then develop a workflow that allows Xiulan and Sofia to review responses to

---

[6]www.hhs.gov/sites/default/files/medicaid-unwinding-letter.pdf

questions and communications generated by the FM. Xiulan and Sofia can edit the responses generated before sending out the final answers, alerts, and notices.

Data engineers develop processes to upload necessary documents to the repository. Information from these documents is embedded into a vector database and passed as context to the FM using the RAG approach.

The data scientists review the FMs and determine the best options for creating alerts or communications using guidelines provided in Chapter 3.

Prompt engineers work on optimizing the prompts necessary to generate the responses to questions as well as the alerts and notices. They also create specific prompts for specific groups to account for regional dialects, cultural norms and values, and more. They work interactively with Xiulan and Sofia, periodically reviewing the output and testing against the established or expected answers for the output.

Outcome and Benefits

Using GenAI to automate the creation of appeals, notices, and communications provides several benefits to Xiulan and Sofia. First, using GenAI, they get the advantage of initial drafts without spending any time. They can then spend time editing the documents as needed to refine the outputs. This especially helps Sofia deal with urgent situations that require getting important public safety information out quickly to the community.

Whether warnings about health risks, natural disasters, or other time-sensitive announcements, the technology enables the information to reach more constituents when it matters most. Xiulan finds that the answers to questions in emails also incorporate regional and cultural customization. She can now quickly review most of the emails and focus on complex situations.

Overall, Xiulian and Sofia are able to get critical alerts and notifications to the public in a timely manner in emergency situations, thus potentially improving the livelihood of constituents.

Additional Considerations

Let's review some additional considerations to ensure the responsible and trustworthy use of this technology:

- **Integration with verification processes**: It is crucial that developers implement robust verification processes to validate the quality and accuracy of the generated content. For example, for questions regarding passports, they can create a human review by passport-related subject matter experts. Alternatively, they can use agents to fact-check against authoritative data sources. Establishing a clear approval workflow and sign-off procedure is essential to maintain the integrity of the information being disseminated.

- **Ongoing monitoring to prevent bias and ensure responsible AI practices**: GenAI systems can inadvertently perpetuate societal biases present in their training data or exhibit other undesirable behaviors. To mitigate these risks, developers should implement ongoing monitoring and evaluation processes to detect and address potential biases or harmful outputs. We discuss these techniques in Chapter 9.

The preceding considerations can help PSOs leverage the power of GenAI for document generation in public communication, alerts, and notices while mitigating potential risks and building public trust. Striking the right balance between automation and human oversight, coupled with a commitment to ethical AI practices, is critical for the success of this technology.

5.3 Image Generation

Similar to document generation, the core of an image generation application involves using an FM to generate required images of different types. However, prompt engineering for an image generation process is more complex. The process is much more iterative and requires careful crafting of prompts. One of the main reasons for this is that image generation FMs, such as Stable Diffusion,[7] have a very limited token size for their prompts in comparison to text generation models; as such, concepts such as RAG for in-context learning used for document generation are generally not applicable. Instead, we use prompt engineering techniques that are specifically geared toward image generation, such as negative prompting. A *negative prompt* is something that you *don't* want to see in image.

Let's look at an example of prompt engineering at work for Stable Diffusion. Let's say you work for a local government office and need to create an image for a campaign encouraging voter participation. You use a simple prompt:

```
People voting happily at a polling station.
```

The result:

---

[7]https://stability.ai/

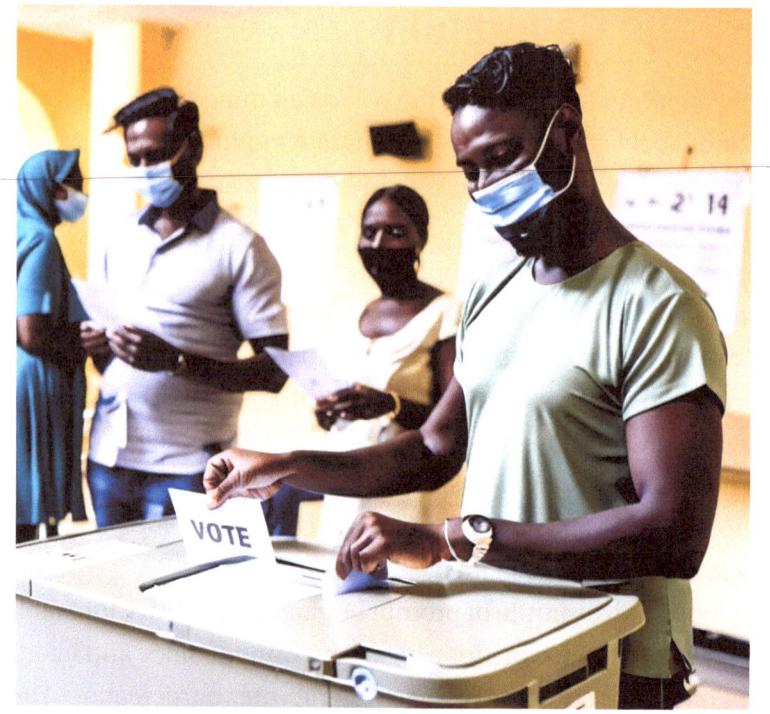

Negative prompt: masks

This image is acceptable, but it has introduced face masks, which aren't required in your locality. Let's add a negative prompt to remove the masks:

```
Negative prompt: masks
```

There you go! By adding `negative prompt: masks`, we tell Stable Diffusion to avoid generating images with those features.

By refining the prompt with more detail, specific adjectives, and references, you can significantly improve the outcome. But even then it might take several tries and adjustments to achieve your ideal image; any solution you choose for image generation needs to incorporate the ability for repeated generation of images, a human review, tuning, and storage for prompts.

5.3.1 High-Level Architecture

Figure 5-2 shows the high-level architecture for an image generation application that extends the base architecture illustrated in Figure 4-4 in Chapter 4, and it accounts for the specific considerations of image generation.

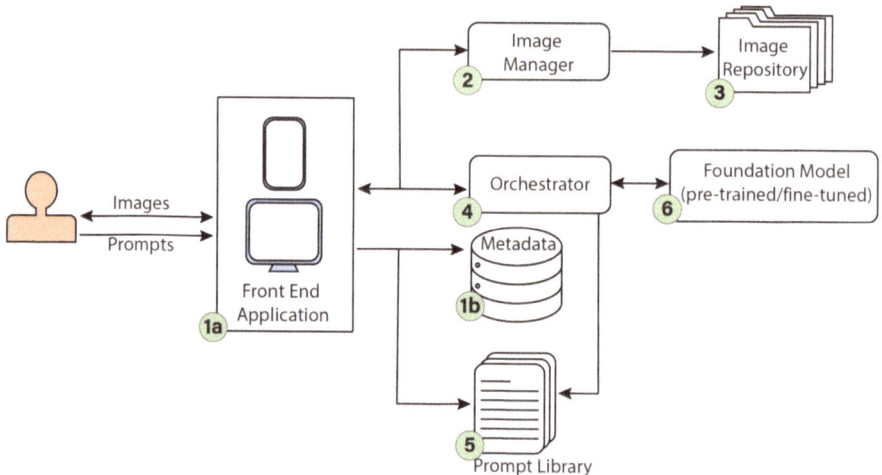

Figure 5-2. *A Solution to Generate Images Using GenAI*

Let's walk through each component of this architecture.

1. **Front-end application**: The user interacts with a front-end application that presents an interactive UI. The user could be a person seeking to generate an image or a reviewer. The application should handle images effectively via capabilities such as editing, cropping, and enhancing the image. Some models can take an image as an input in addition to or instead of a prompt, so the user should also be able to upload an image. Once the initial image is generated by the FM, it is important that a user

has the ability to edit the image as necessary. For
example, a user should be able to enhance or modify
the initial image generated and submit it again
for further refinement. The front-end application
should also have the ability to save prompts selected
by the user into a library. The prompts should be
versioned and be readily available for the user as
needed. Users should be able to share prompts to
enable collaboration and improvement.

2. **Image manager**: The front-end interfaces with an
   image manager that is responsible for retrieving and
   storing images in a repository. It should have the
   ability to store versions of images in various formats.

3. **Repository**: The repository stores the generated
   images and interfaces with the image manager.

4. **Orchestrator**: The front-end application invokes an
   orchestrator (as described in the previous chapter)
   that handles the heavy lifting by interacting with the
   FM to generate images.

5. **Prompt library**: Prompts are saved in this library.
   Users can save, edit, and modify the prompts from
   this library using the front-end application. You
   can accomplish this integration directly or use the
   orchestrator.

6. **FM**: This can be an FM of your choice, either the
   base model or fine-tuned.

Now that we've reviewed the main components of an image generation
application, let's discuss some additional aspects we need to consider.

5.3.2 Key Considerations

- **Collaboration**: Both internal and external collaboration is essential to develop and improve the prompts used. Making the entire image generation process more accessible, user-friendly, and efficient ultimately enhances the quality of application and their output.

- **Ethical considerations**: PSOs have to be mindful of potential misuse. Implement safeguards and fact-checking mechanisms to ensure that generated images are handled responsibly and adhere to data privacy regulations. Consider using technology such as watermarks to differentiate images generated by FMs.

- **Copyright and intellectual property**: Understand and comply with copyright laws regarding generated images, especially when using existing content as training data. Clearly define licensing terms and usage rights for generated images, considering public access and potential commercial applications.

- **Social impact**: Image generation when used for right use cases can yield tremendous benefits in the public sector. However, before embarking on image generation projects, you should carefully evaluate the cost-effectiveness of image generation compared to traditional methods. The process can be time-consuming and expensive and should be used only when there is a clear gap or requirement. Next, we discuss two such use cases to serve as examples.

Next, we explore two different use cases of image generation in the public sector. As we did for document generation, we discuss the different roles involved in implementing the solution.

5.3.3 Image Generation Use Case 1: Synthetic Image Generation for Medical Research

Problem

Medical research is critical for finding potential cures to many diseases and conditions. Successful research is heavily dependent on obtaining medical images for tasks such as disease simulation, epidemiology, simulation of molecular structures, and outbreak predictions. However, in many cases, medical images are not available and/or labeled for medical research.

Let's take the example of a hypothetical PSO involved in medical research. Shirley is a medical researcher focusing on the impacts of COVID. In order to study the different potential impacts, she needs to obtain as many radiology images as possible of infections with different strains of the virus. However, due to the rapidly evolving nature of COVID, obtaining images has been very challenging.

Solution

Traditional techniques such as mathematical and statistical models can serve these research needs well, but the advent of GenAI offers a significant leap in the ability to create realistic images in a controlled and efficient manner. Images created using GenAI are visually realistic and closely resemble real data. You can fine-tune the generation process, controlling specific features and pathologies.

Therefore, the CIO offers to work with Shirley on an initiative to use GenAI for synthetic images. Her team implements the high-level architecture in Figure 5-2 to implement an image generation solution for medical research.

The developers understand that the front end plays a critical role in the architecture. They design a front end that caters to Shirley's needs, specifically with options for selecting specific modalities (MRI, CT, and so on), anatomical regions, and pathologies or variations. They build a repository that can securely store diverse medical image formats, including real data, annotations, and generated images. They also make sure that the front end allows for management of repository metadata such as diagnosis, conditions associated with each image for efficient search, filtering, and analysis.

Uploading a few real medical images as references is crucial, as that allows Shirley to take existing images, and make tailored and precise modifications needed for her research. The developers also integrate the front end with editing tools that allow precise modification of generated images to match the exact anatomical features and variations that Shirley desires. They include a tool that applies post-processing steps (denoising, anatomical consistency checks) to ensure quality.

The data scientists review the different text-to-image models and determine the best FMs for this use case using guidelines provided in Chapter 3. They also coordinate with ML engineers to fine-tune the models specifically for the specific variations and conditions of COVID.

Prompt engineers work iteratively with Shirley to develop multiple benchmark prompts using specific research questions and disease models. They save those in the prompt library for reuse and collaboration with other medical researchers. The library includes annotations with metadata (modality, pathology, and so on) to facilitate search and understanding.

Outcome and Benefits

Shirley is now able to create several variations of images of specific conditions using natural language prompts. Potential applications are vast, spanning drug development, medical imaging analysis, clinical education, and public health research. Imagine training doctors on rare medical cases in simulated environments or studying disease outbreaks without real-world risks! In addition, this solution can also be used for generating realistic visual aids to explain medical conditions and treatment options to patients.

Additional Considerations

- **Integration with medical devices**: Many sources of images are medical devices with proprietary formats. Data engineers need to work with these vendors to allow exports of data in a format acceptable for the FM.

- **Data privacy and security**: Security specialists need to make sure that robust security measures are in place to protect sensitive patient information and to ensure compliance with medical data privacy regulations. We discuss some of these measures in Chapter 9.

- **Validation and testing**: Given the purely generative nature of this solution, rigorous validation protocols must be in place to assess the quality, realism, and anatomical accuracy of generated images. Shirley needs to review the images being created with her colleagues to validate the accuracy.

- **Documentation and transparency**: The team should comprehensively document the image generation process, including model architectures, training data, and limitations. ML engineers should incorporate techniques such as watermarks (Chapter 3) to indicate that the images are AI generated. This promotes transparency and reproducibility in research.

5.3.4 Image Generation Use Case 2: Education

Problem

Traditional educational resources often lack engaging visuals, diversity, and accessibility. Jie, an instructor in a PSO involved with education, finds that generating high-quality images for specific needs can be time-consuming and expensive. Additionally, representing diverse cultures and historical periods through visuals can be challenging with real-world limitations.

Solution

GenAI offers the ability to create synthetic images on demand. The IT director determines that she has the budget to help Jie. She forms a team that implements a solution based on Figure 5-2.

Developers create a front-end application that allows Jie to choose the type, topic, style, and cultural context of the image that she wants. They also provide the capability to upload the baseline images that Jie may want to alter. They integrate the front end with image viewing and editing tools so that once an image is generated, Jie can further tailor the image as required to suit her exact needs. They integrate the front end with a repository to store generated images with metadata for easy search and retrieval.

Data scientists evaluate and select the FM most suited for the use case using guidelines provided in Chapter 3. ML engineers implement watermarks (Chapter 3) to show that the source of the image is GenAI.

Prompt engineers work with Jie to develop, catalog, and organize prompts by subject, age group, and accessibility needs. They use negative prompting to make sure that images do not contain sensitive material. In most cases, they are able to generate the images Jie desires without any fine-tuning. The team informs Jie that the option for fine-tuning of the FMs exists if the results do not meet her needs. Jie approves the implementation since the results meet her needs for most subjects. She will work with the IT director on a future project that incorporates fine-tuning.

Outcome and Benefits

The benefits of using GenAI to generate images in education are multifold across various fields and subjects. By prioritizing her needs, Jie is now able to generate realistic simulations of many different subjects very quickly. She finds that she can dramatically improve student engagement and foster interest by using compelling visuals. She is also able to tailor learning materials with the appropriate level of complexity and representation for students with specific needs.

Additional Considerations

- **Checks for accuracy**: Considering that these tools are being used for education, Jie needs to ensure there are checks in place for ensuring the images are accurate and appropriate for education. She can work with the team to incorporate workflows for reviews, audits, and checks with other qualified staff.

- **Data privacy and security**: Security specialists need to make sure that there are checks in place to ensure that student data privacy regulations are not violated. We discuss some of these checks in Chapter 9.

- **Transparency**: Jie should disclose the use of GenAI-created images and educate students on responsible digital literacy.

By harnessing the power of GenAI, public education can become more engaging, diverse, and accessible. Responsible development and ethical considerations are crucial to maximize the benefits while mitigating potential risks. Imagine history textbooks coming alive with culturally sensitive visuals and science lessons are enhanced with interactive simulations. GenAI has the potential to revolutionize the education industry and create positive experiences for us all.

5.4 Code Generation

As discussed earlier in the chapter, PSOs struggle with resource limitations and the need for efficient, secure, and accessible software solutions. From a broader perspective, content generation using GenAI can help with creating a lot of artifacts used within a Software Development Life Cycle (SDLC). This includes creating project plans, architecture and design documents, test plans, technical documentation, and…code!

Code generation is a special case, because of the complexity involved in programming. Traditionally, programming was perceived as a hard skill that required specialized resources and staff. However, GenAI-based code generation tools are now able to create code for many use cases. There are a plethora of specialized tools that use FMs for providing the capability of generating code, some of which are included in Appendix C. In addition, a number of open source, publicly available, or proprietary FMs also offer code generation capabilities; some are listed in Appendix A. The primary benefit of these tools is to allow programmers to focus on complex problem-solving and logical thinking and use code generation tools for tasks that are considered repetitive and tedious. So now, Mary, the IT director, can have her team pay attention to the most complex and exciting problems. This in turn improves her team's morale, productivity, and the overall quality of code.

Let's look at all the different tasks that code generation tools can support:

- **Code completion and auto-suggestion**: These tools analyze your existing code and suggest relevant completions for functions, variables, and class names. This functionality helps developers write code faster and with fewer errors, improving productivity.

- **Code snippet generation**: Based on your input (natural language description, existing code, and so on), these tools generate complete code snippets for specific functionalities. This can be helpful for tasks such as creating common data structures, implementing algorithms, or integrating with APIs.

- **Code analysis and debugging**: You can use GenAI tools to analyze your code and flag potential issues such as logical flaws, syntax errors, or inefficient coding practices. These tools can also propose potential fixes or enhancements. This can be a huge time-saver compared to manually sifting through lines of code.

- **Boilerplate code generation**: Many development tasks involve repetitive code for setting up common functionalities including database connections, data transformations, error handling, or user authentication. Code generation tools automate this process, saving developers time and effort.

- **Unit test generation**: These tools can automatically generate unit tests based on your existing code, ensuring basic functionality and reducing the manual effort involved in testing. This helps improve code quality and maintainability.

- **Refactoring and code improvement**: Some tools can analyze your code and suggest improvements such as renaming variables, refactoring code structure, or optimizing algorithms. This helps maintain clean and efficient code, improving readability and maintainability.

- **Data processing and analysis**: Code generation tools can be used to automate data cleaning, transformation, and analysis tasks based on specific requirements. This can save data scientists and analysts time on repetitive coding tasks, allowing them to focus on more complex analysis.

- **User interface (UI) and API generation**: Some tools can generate basic UI components or API endpoints based on user specifications or existing data models. This can be helpful for prototyping or quickly building simple interfaces, although complex UIs might still require manual design and development.

- **Secure code generation**: Some tools can be used to ensure that the code follows secure coding practices, thereby reducing vulnerabilities and improving security posture. This also reduces the risk of introducing expensive errors later in the development life cycle.

5.4.1 Code Generation Evaluation Framework

With so many tools and so many different types of functionalities, evaluating these tools is a challenge. We provide a framework in Appendix G to evaluate these tools. Needless to say, requirements specific to the organization should also be incorporated.

At the present time, we advocate considering these tools more as productivity boosters and aids for specific tasks, rather than a means for generating entire applications or complex functionalities from scratch. However, given the incredible rate of innovation in this space, that may happen sooner than we think!

5.5 Conclusion

In this chapter, we discussed different use cases related to content generation in the public sector. We discussed using document generation, image generation, and code generation for a variety of tasks and discussed specific use cases related to these tasks.

Remember, we have only scratched the surface of GenAI's content generation potential. As the technology evolves and PSOs gain experience, we can expect more innovative applications to emerge.

As discussed in Chapter 3, careful and responsible deployment of these technologies is very important. Taking measures to mitigate potential biases, transparency in decision-making, and robust security measures can unlock tremendous benefits that will ultimately translate to improvements to society. In the next chapter, we will discuss how GenAI-powered chatbots can enable the effective communication, constituent engagement, and productivity of PSOs.

CHAPTER 6

GenAI-Powered Chatbots

In Chapter 5, we discussed several Generative AI (GenAI) use cases for content generation in public sector organizations (PSOs). We covered document generation, image generation, and code generation, each with slightly different architectures, implementation considerations, and use case examples. In this chapter, we'll take a deeper look at another widely adopted use case: chatbots.

We have all heard of chatbots. Similar to the other industry verticals, PSOs have been adopting chatbots to improve customer service and responsiveness. Chatbots offer automated customer service 24/7 and can provide instant responses to frequently asked questions without the need for human agents. As an example, a powerful chatbot can direct constituents to the correct process, provide links to relevant online resources, or even take simple actions such as changing an address, all within a conversational interface. Using chatbots for routine inquiries and tasks frees up staff members to focus on more complex tasks that require human analysis, judgment, and subject matter expertise. Given the potential benefits, the adoption of chatbots has been on the rise.

However, the advent of GenAI dramatically expands the scope and potential of chatbots. To understand further, we need to dive a bit deeper into the differences between traditional and GenAI-based chatbots.

6.1 Differences Between Traditional and GenAI-Based Techniques

As we touched upon briefly in Chapter 2, there are a few differences between traditional and GenAI-based chatbots.

The traditional techniques utilized machine learning–based models to translate utterances to intent using supervised learning methods; see Chapter 1 for a discussion of supervised learning. The idea is that people communicate in many ways – *utterances* – for the same underlying *intent*. The intent is mapped to a set of corresponding answers. So, once a model can predict an intent, then the corresponding answer is provided.

The problem with this approach is twofold. First, in order for the chatbot to work well, it needs to be trained with a supervised learning approach. This includes collecting a large corpus of sample conversations or utterances, labeling them with their corresponding intents or desired actions/responses, and then training them using an ML algorithm; this training can be time-consuming. Second, even after training, the ability of the chatbot to handle complex questions is limited because the answers are scripted and mostly static.

GenAI-based chatbots use Foundation Models (FMs). The FMs take in the chat inputs as prompts and generate outputs. This not only overcomes the preceding issues but also provides a number of additional advantages including

- **Open-ended conversation**: Since FMs are trained on vast amounts of data, they can generate human-like responses dynamically. This allows for more open-ended and natural conversations that go beyond predefined scripted responses.

- **Context and intent understanding**: FMs can understand the context better and underlying intent behind user utterances, enabling more nuanced and appropriate responses.

- **Adaptability**: Since FMs are trained on diverse data, they can adapt to new domains and topics relatively easily, reducing the need for extensive data collection and annotation for every utterance.

- **Personalization**: Many FMs can handle multiple languages. They can also take on different roles and adapt their personality to match individual user preferences, leading to more engaging and personalized conversations.

Because of these advantages, the potential for the use of GenAI-powered chatbots is vastly expanded compared to traditional methods.

6.2 Broad Areas of GenAI Chatbot Usage and Challenges

There are two broad areas in which GenAI chatbots can be used in the public sector:

1. **Improving constituent services**: Chatbots can offer tailored guidance and assistance based on individual needs and inquiries, catering to diverse populations through multilingual capabilities. They can also anticipate constituent needs and proactively offer relevant information or services, improving overall satisfaction. As an example, a GenAI-powered chatbot can help the constituent to change their address automatically so that they don't get rejected for benefit re-enrollment due to incorrect address.[1]

---

[1] https://aws.amazon.com/blogs/publicsector/automating-returned-mail-keep-members-enrolled-during-medicaid-unwinding/

2. **Enhancing operational efficiency**: Chatbots can guide employees through internal workflows and procedures, reducing paperwork and wait times. As an example, they can help the workforce with human resources (HR)–related, IT-related, or cyber-related policies or help them with other frequently asked questions. The GenAI-powered chatbots can help improve organizational processes and decision-making. They can provide clear and consistent information internally, promoting trust and collaboration.

Overall, GenAI chatbots offer PSOs a powerful tool to improve both constituent service delivery and internal operations. However, there are some challenges with GenAI-based chatbots; these include

- **Potential for inaccurate or biased responses**: Since FMs generate responses dynamically, there is a risk of producing inaccurate, biased, or inappropriate responses called *hallucinations*.

- **Consistency and coherence issues**: Maintaining consistent and coherent responses across multiple turns of conversation can be challenging, particularly for longer conversations.

In the next section, we discuss a high-level architecture for building GenAI-based chatbots that address the preceding challenges.

6.3 High-Level Architecture

As mentioned earlier, GenAI chatbots have their own set of challenges. One of the biggest challenges is the potential for inaccurate responses and hallucinations. As discussed in Chapter 4, some of the best ways to overcome this limitation is through the use of prompt engineering and Retrieval-Augmented Generation (RAG) techniques (Section 4.6.1).

To review the concept of RAG, it helps retrieve and use external knowledge to generate better responses that are grounded in facts rather than purely relying on the pre-trained FM.

The RAG process includes two phases: the initial data processing phase and the execution phase. The initial data processing phase converts domain knowledge data into embeddings and stores them in a vector database. The execution phase consists of three steps:

1. **Retrieval**: Given a prompt, retrieve relevant information from the vector database.

2. **Augmentation**: Retrieved information goes through post-processing, and the final contextual information is used to enrich the prompt.

3. **Generation**: The FM generates the final response using enriched prompts.

With this concept in mind, we use an example to walk through how a GenAI-powered chatbot would work. Take the case of John, the caseworker in a PSO. John is not comfortable with technology. He frequently needs to call or chat with the IT support team. Mary, the IT director, decides to implement a GenAI chatbot for IT support that can help people such as John. Her team builds the chatbot using the RAG architecture with documents that contain information about IT systems.

Here's what happens when John uses the chatbot:

1. John enters an IT support–related question in the chatbot.

2. Using RAG, relevant sections of information are retrieved from the documents (retrieval).

3. The preceding information is appended to John's question (augmentation).

4. The question and additional information are provided to the FM. The FM generates a response, which is then provided back to John.

The preceding technique ensures that John gets the most relevant answer to his question. Now let's take a look at a high-level architecture of a chatbot that incorporates this workflow, as illustrated in Figure 6-1.

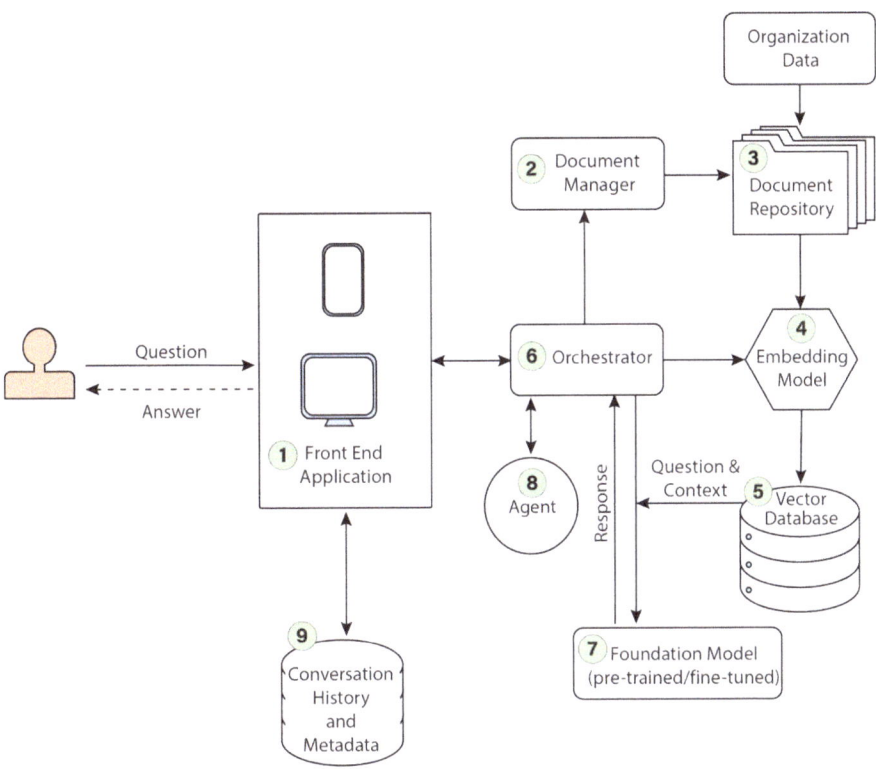

Figure 6-1. *GenAI-Powered Chatbot High-Level Architecture*

The architecture depicted in Figure 6-1 not only implements the workflow described earlier but also includes additional features that address requirements specific to chatbots. The components are as follows:

1. **Front-end application**: The front-end application provides an interactive UI. The UI can be designed in different formats, as a question and answer site or a pop-up window on a website. Users enter their questions and receive answers in the front-end application in an interactive manner. The answers could also include links to source documents. The front-end application provides other functionalities,

such as collecting feedback on the response, options for commonly seen follow-up questions, or clickable buttons for further tasks or even allowing users to upload documents.

2. **Document manager**: The document manager interfaces with a document repository. It enables the system to store, add, delete, modify, organize, and sort documents.

3. **Document repository**: A document repository stores source documents that would be helpful for the chat. This includes unstructured documents in various formats such as PDF or spreadsheet and so on. It also integrates resources such as web pages, code repositories, and structured databases. The document repository is updated regularly.

4. **Embedding model**: An embedding model converts document content from the document repository into numerical representations called *embeddings*; see Chapter 1 for an explanation of embeddings. The embedding model also converts user questions into embeddings. Note that the embedding model can handle both text and images, which are often valuable sources of information in chatbot conversations. This versatility ensures that the chatbot can leverage both textual and visual data to provide the most comprehensive and accurate responses.

5. **Vector database**: A vector database is used to store document's embeddings and retrieve context relevant to the user question by using a machine

learning technique called similarity search. Most vector databases available today support similarity search; see Appendix B for a list of available vector databases.

6. **Orchestrator**: The front-end application invokes an orchestrator that does the heavy lifting with prompt chaining and managing interactions with different components. We discussed prompt chaining in Chapter 4. In this example, the orchestrator calls the embedding model to convert a user question into embeddings, uses the retrieved context to enrich the original user question, interacts with FM to generate the response, and sends it back to the front-end application while triggering some other tasks via agents. Designing an efficient orchestrator is critical to the success of this solution.

7. **Foundation Model**: This can be an FM of your choice, either a base model or a fine-tuned model. As mentioned before, fine-tuning may still be required if you have specific domain nomenclature such as healthcare, education, or transportation. FMs in common chatbot use cases are either text-to-text models that accept text as input or multimodality models that accept text together with some other modality of data such as images and even audio and video.

8. **Agent**: As described in Chapter 4, an agent enables integration with internal or external tools to perform certain actions to expand a chatbot or enterprise search engine's functionality and achieve complex

business tasks. Let's take the same example of a chatbot for IT support. As opposed to just answering questions about IT support, an agent can also perform functions such as creating a ticket in the troubleshooting or help desk system.

9. **Conversation history and metadata**: This is a critical component of the overall architecture. Chatbots need to understand the context of the conversation to provide relevant and coherent responses. However, as discussed in Chapter 1, FMs are stateless. In other words, they simply respond to a prompt with a response, without taking into account previous exchanges.

Most conversations on the other hand involve multiple back-and-forth exchanges, where the chatbot needs to remember and reference previous messages to maintain a coherent dialogue.

To understand how to overcome this problem, let's take the same example of the IT support chatbot. John types in a question and receives an answer. When John types in his next question, the previous question and answer are appended to the current question. This continues till the end of the conversation.

This approach partially addresses the challenge mentioned earlier with consistency and coherence. By having access to the conversation history, the chatbot can maintain the continuity and flow of the conversation, avoiding abrupt topic changes or disconnected responses and creating a more natural and seamless user experience.

Front-end application techniques to maintain conversation history can involve mechanisms such as browser cookies or local storage on the user device. However, an enterprise-level chatbot uses more robust solutions such as databases or in-memory caches. Now, there are some problems with this, such as the increasing length of the context, costs, handling changes in topics, etc. We discuss some of these in the next section on key considerations.

6.4 Key Considerations

In this section, we address some of the nuances and challenges that arise when building a chatbot, as follows:

- Data privacy and content filtering

- Document chunking

- Handling complex documents

- Graph databases

- Memory management

- Context switching

- Fine-tuning

- Integration with the user interface (UI)

- Feedback collection

- Trigger other actions

- Performance

- Cost

6.4.1 Data Privacy and Content Filtering

Data privacy and content filtering are absolutely crucial for public sector chatbots due to the sensitive nature of the information they handle and the potential for misuse. This is especially true for any constituent facing chatbots. Let's review some of these considerations in detail:

- **Data privacy**: Imagine a public sector chatbot assisting Jane, a constituent, with healthcare inquiries. This chatbot might collect sensitive personal data such as medical history, insurance details, or prescription information. Implementing robust data privacy measures such as secure storage, encryption, and access control mechanisms is paramount to safeguard this information from unauthorized access, breaches, or misuse. Additionally, transparency regarding data collection practices and compliance with regulations, such as Health Insurance Portability and Accountability Act (HIPAA), is essential for building trust and encouraging constituents to utilize these services without privacy concerns.

- **Content filtering**: Public sector chatbots are susceptible to malicious actors attempting to inject harmful content such as hate speech, misinformation, or spam. Imagine the chatbot designed to answer Jane's questions being manipulated to spread false information about vaccines. Implementing robust content filtering mechanisms is vital to prevent the spread of such harmful content and maintain a safe and secure environment for constituents interacting with the chatbot.

Therefore, prioritizing data privacy and content filtering is essential for public sector chatbots to operate ethically and responsibly and maintain the trust of constituents as well as internal employees. In Chapter 3, we discussed how PSOs can ensure the safe and secure handling of sensitive information using guardrails. In Chapter 9, we will dive deeper into enhancing data privacy and security for various components of a GenAI application.

6.4.2 Document Chunking

When converting external documents into embeddings and storing them in a vector database, the first step involves breaking the content down into smaller segments; this is known as *chunking*. Chunking breaks down large documents into smaller, manageable segments. This allows for a more granular match between the user query and specific parts of the text, leading to more precise and relevant retrieval of information.

For chunking to work properly, determining the optimal chunk size is essential. If the chunk is too small, you risk missing relevant information or generating incoherent responses. If it is too large, it can overwhelm the system with too much information which leads to increased costs.

Take a step back to understand why the cost may increase, which may happen, especially if you use FMs from cloud-based providers. Most providers charge an amount based on the number of tokens processed by the FM; see Chapter 1 for an explanation of tokens. If you feed in a large chunk of text, you will incur higher costs due to the larger number of tokens.

Going back to chunking, in addition to chunk size, there are other factors that need to be considered. These include the nature of the content, the embedding model and its optimal block size, the expected length and complexity of front-end user queries, and how the retrieved results will be used in the augmentation step. While a detailed discussion on chunking is out of scope for this book, we review some of the key chunking strategies for consideration:

- **Fixed-size chunking**: This is a simple and efficient method, ideal for situations where consistent chunk size is crucial. For example, a chatbot built for legal may benefit from consistent chunking. In this case, the fixed size ensures the retrieval of specific regulations with predictable lengths, making it easier to pinpoint the relevant sections within the retrieved information.

- **Structure-aware splitting**: This strategy preserves the natural structure of documents, making it suitable for content organized into sections or paragraphs. For example, splitting a permit application document by sections (eligibility criteria, application process, fees) provides a more structured response.

- **Recursive structure-aware splitting**: This hybrid approach balances fixed-size chunks with linguistic boundaries. It is useful when dealing with content with different section lengths but still requires some structure preservation. As an example, this method can be used in a social security benefit enrollment document. This approach helps in retrieving relevant sections while still maintaining manageable chunk sizes.

- **Content-aware splitting**: This strategy excels in handling structured documents with distinct content types such as HTML or LaTex, such as a public website.

Content-aware splitting ensures separate chunks for text content, images, and embedded tables, leading to accurate retrieval and presentation.

- **Natural language processing (NLP) chunking**: This advanced approach leverages NLP techniques to identify topic shifts within text. While this approach is powerful, it requires sophisticated technology. This method would be ideal for handling complex inquiries that span multiple topics within a single document, such as a chatbot providing a comprehensive overview of environmental regulations across different sectors.

Ultimately, the optimal chunking strategy depends on the specific needs of the task and the nature of the documents processed. Open source packages such as LangChain[2] and LlamaIndex[3] have these chunking strategies built in as classes and functions. Cloud vendors such as AWS, Azure, and Google usually provide API services that integrate certain strategies to chunk documents for generating embeddings as part of the document ingestion step.

6.4.3 Handling Complex Documents

Real-world documents contain text, tables, and images to convey complex content; we discussed a multimodal FM in Chapter 1. A multimodal RAG pipeline helps with both images and text to be stored in the vector database for information retrieval.

Each modality (text, images) has its own unique characteristics. The challenge lies in effectively representing and processing information from these diverse sources within a unified framework. Effectively managing

---

[2] www.langchain.com/
[3] www.llamaindex.ai/

the relationships and connections between information across different modalities is crucial for accurate retrieval and response generation. Here are some approaches to address these challenges:[4]

- **Embedding all modalities into the same vector database**: This approach leverages the existing text-only RAG setup and replaces the embedding model with one capable of handling multiple modalities. For example, models such as Contrastive Language–Image Pre-training (CLIP) can encode both text and images into the same vector space, allowing them to be stored and retrieved together. So an inquiry about a specific government program might involve retrieving relevant text and images from regulations.

- **Grounding modalities into a primary modality**: This approach selects a primary modality (most often text) and transforms all other modalities into that format. For example, if the primary focus is text-based question and answer, images can be preprocessed by generating textual descriptions and metadata using an MLLM. These are then embedded and stored alongside the original text. Images can be stored separately for potential use during response generation.

For example, a constituent, Jane, wants to ask a chatbot about public transportation schedules to go to her benefits eligibility interview. In this case, the bus routes are converted into textual descriptions by an MLLM and embedded in the vector database. The question is then mapped to these descriptions for retrieval of potential bus routes. The corresponding images of matched bus routes are then presented back to Jane.

---

[4] https://developer.nvidia.com/blog/an-easy-introduction-to-multimodal-retrieval-augmented-generation/

- **Separate vector databases for different modalities**: This approach, also known as *rank-rerank*, uses the same embedding approach as the previous one but involves storing vectors for text and images separately. During retrieval, each modality is queried independently to retrieve top-ranked information. A dedicated multimodal reranker then analyzes the returned information to provide the most relevant overall response.

Jane has a question about healthcare insurance plans. In this case, the question is mapped to both the image database containing images of various plans and vendors as well as a text database containing healthcare regulations, FAQs, and guidelines. The retrieved results are combined and ranked, and the most relevant information is used to generate the response. For example, the final response could display the guidelines as well as an image of the insurance plan, making it easy for Jane to identify the correct plan.

6.4.4 Graph Databases

Recently, some researchers and developers have started using graph databases to replace vector databases in RAG solutions.[5] Compared to vector databases that partition and index data using embedding model encoded vectors, graph databases construct a knowledge graph from extracted entity relationships within the text. Graph databases can make retrievals more concise by fetching more relevant data. However, this area is still under research as of this time, and a solution based on graph databases may have latency and performance implications.

---

[5] https://neo4j.com/developer-blog/knowledge-graphs-llms-multi-hop-question-answering/#:~:text=Information%20spanning%20multiple%20documents

6.4.5 Memory Management

As discussed earlier, chatbots lack inherent memory. This means that each user query is treated independently, without considering past interactions. To overcome this limitation and to create a more natural and helpful user experience, developers need to implement various techniques including

- **Total recall**: Let's imagine a constituent, Jane, asks a government agency chatbot about her eligibility for a specific social program. In this approach, the entire history of Jane's conversation including each question and answer is appended to every new query. While technically straightforward, this method can quickly lead to performance issues as the conversation lengthens and the data volume grows. Additionally, the combined query and context might exceed the processing capacity of the FM powering the chatbot.

- **Sliding window**: Similar to the total recall approach, the sliding window technique includes past interactions. However, it focuses on the most recent messages within a predefined window. This is analogous to Jane interacting with a government representative, where the focus is on the current topic and immediate context. Older messages are gradually forgotten as newer ones are added. This approach addresses the performance concerns of total recall but risks losing crucial context from earlier parts of the conversation, potentially leading to inaccurate responses if the user refers back to something outside the window.

- **Summarization**: This approach condenses the chat history into key points, similar to how Jane might summarize her conversation with a government agency representative at the end of the interaction. This reduces the data size while preserving essential information for a coherent conversation. However, summarization can lead to the loss of specific details that might be relevant later.

- **Vector embeddings**: This technique leverages the existing vector database used in RAG models to store historical conversation embeddings. Similar to how Jane's previous inquiries about public services might influence their current request, this method identifies relevant past interactions based on their similarity to the current query. This approach offers dynamic adaptation based on the conversation flow but requires more complex implementation compared to other methods.

Choosing the most suitable memory technique depends on the specific needs of the application and the desired balance between context retention, performance, and implementation complexity.

6.4.6 Context Switching

The ability for a chatbot or enterprise search engine to switch between contexts is critical for a smooth conversation and a positive user experience. *Context switch* occurs when a user is talking about one topic and changes to another subject midway through a conversation, which would be very confusing for the application. By default, GenAI-powered chatbots inherently don't have memory; hence, they deal with context switching much better compared to traditional chatbots. However, the situation changes once we start using some of the memory techniques as explained earlier. In this case, because the context does contain memory in the form of previous history, changing the context abruptly may cause disruption in the user experience.

One method to overcome this issue is to set some escape actions in the user interface. For example, let's say Jane is done asking about social programs and now wants to know about applying for a permit. She can press a restart button that clears the entire chat history.

More advanced techniques include methods such as Named Entity Recognition (NER). In this case, using prompt engineering, you use an FM to identify and track specific entities such as names, locations, and dates throughout the conversation. If the entities start changing, then the chatbot could pose a question to ask if the user intends to change the topic. In the previous example, when Jane is asking questions about social programs, the chatbot captures the key entity as a social program. When she switches to ask questions about a permit, then the key entity for the question is identified as the permit. Since that differs from the previous key entity, the chatbot asks Jane if she no longer wants to talk about social programs. If she says yes, then the chat history is reset.

6.4.7 Fine-Tuning

You may ask this question: Is fine-tuning needed for GenAI chatbots that use RAG?

In most cases, the answer is no. The beauty of RAG lies in its ability to adapt to dynamic data and external information retrieved from different document stores. This adaptation often negates the need for extensive fine-tuning of the FM itself. However, there are specific situations where fine-tuning the FM might be beneficial:

- **Specialized terminology**: If the public sector domain involves unique nomenclature or terminology not present in the FM's pre-training data, the FM might struggle to accurately generate responses using that terminology. This is true even with RAG's retrieval capabilities. In such cases, fine-tuning the FM on a dataset specifically containing relevant domain-specific language can be helpful. Let's imagine a chatbot assisting constituents such as Jane with complex legal inquiries. Legal terminology might not be well represented in the FM's pre-training data. Fine-tuning the FM on a dataset of legal documents and case studies could improve its ability to generate accurate responses tailored to legal inquiries.

- **Brand voice and consistency**: While RAG helps the FM adapt to the context of the conversation, it might not fully capture the PSO's specific brand voice and communication style. If maintaining a consistent brand voice is crucial, fine-tuning the FM on a dataset of internal documents and communication materials can be valuable. Let's use the same example of Jane, a constituent, interacting with a PSO's chatbot. Now, Jane has often interacted directly with staff and members of this agency. So she may expect a similar tone and approach when dealing with the chatbot. Fine-tuning the FM on internal communication guidelines and public service announcements can help achieve this brand consistency.

It is important to weigh the potential benefits of fine-tuning against the additional development effort required. In many public sector applications, the adaptability provided by RAG might be sufficient for effective chatbot and search engine functionality. However, for situations requiring highly specialized language or strict brand adherence, fine-tuning the FM can be a valuable tool for further enhancing the user experience.

6.4.8 Integration with the UI

The chatbot UI is an important component with which constituents and internal employees interact. It encompasses everything from the text displayed on screen to the buttons and menus used to navigate the application and access services. Here are some key points to consider when developing the UI:

- **Intuitive and user-friendly**: The UI should be easy to navigate for users of all backgrounds. Using the earlier example of Jane, the chatbot assisting with social security inquiries should be clear and simple, with easily identifiable buttons for tasks such as checking eligibility, submitting applications, or tracking the status of claims.

- **Visually appealing and brand consistent**: A visually appealing interface with the PSO's brand colors and logo fosters trust and familiarity. For example, a public health department chatbot might utilize a UI that reflects the department's official branding, making it instantly recognizable to constituents.

- **Conversational flow and user feedback**: The UI should clearly indicate when the chatbot is processing information vs. waiting for user input. This can be achieved through visual cues such as typing indicators or animated avatars. Additionally, providing options for feedback buttons or satisfaction ratings allows users to gauge the chatbot's performance and offer valuable insights for improvement. So, business users should be able to provide feedback if the IT support chatbot doesn't answer their question as clearly as expected.

- **Technical integration**: Seamless integration between the UI and the chatbot's back end is crucial. This ensures smooth information exchange and data compatibility between the two systems. The API and data formats used for user queries and chatbot responses need to function flawlessly on both sides. For example, if Jane asks a question about the benefits amount she is going to receive for a social program, the response amount retrieved by the API from a database should be correctly handled and presented in a numerical format in the UI.

- **Thorough testing and error handling**: Rigorous testing of the integrated system is essential to guarantee smooth conversation flow and to address potential issues. Corner cases and errors should be gracefully handled to avoid confusing users and ensure a positive user experience. Imagine a scenario where Jane encounters an unexpected error message while using a chatbot for tax filing assistance. A well-designed UI would provide clear instructions or alternative options to guide Jane through the issue without causing frustration.

By prioritizing these factors, PSOs can create effective and user-friendly chatbot UIs that enhance constituent engagement as well as internal operations.

6.4.9 Triggering Other Actions

The capabilities of GenAI-powered chatbots can be further enhanced by triggering actions beyond simply providing information. This automation adds a new layer of functionality and streamlines service delivery for users.

Imagine a government chatbot assisting John, the agency caseworker, with replacing his desktop's monitor. Beyond answering questions about renewal procedures, the chatbot can be integrated with the PSO's procurement system and IT help desk to

- **Schedule appointments**: The chatbot can analyze John's request and, if applicable, trigger an action to schedule an appointment with the IT help desk.

- **Submit request**: Based on John's input, the chatbot can initiate the process of submitting the request for the desktop monitor electronically.

These actions are accomplished using the agent component of the high-level architecture in Figure 6-1. We explained how an agent works in detail in Chapter 4.

To trigger actions, the agent needs to be integrated with internal or external tools. This integration is typically achieved using relevant Application Programming Interfaces (APIs) and software development kits (SDKs) for communication. Implementing appropriate error handling and fallback mechanisms is crucial. So, let's say that the desktop monitor procurement action fails due to technical issues or unforeseen circumstances, John receives clear communication and alternative options.

By incorporating action-triggering capabilities, public sector chatbots can evolve beyond the simple task of providing information. This includes capabilities to streamline service delivery and enhance the overall service experience.

6.4.10 Performance Considerations

In this information age, constituents as well as internal users expect prompt and efficient responses from the PSO sources, and a chatbot is no exception. Slow response times can lead to frustration, a decline in user satisfaction, and potentially lost opportunities to serve constituents effectively. Therefore, ensuring performance of the application is crucial to the success of a chatbot. In Chapter 9, we will discuss the various methods of enhancing the performance of a GenAI application.

6.4.11 Cost Considerations

PSOs typically operate with limited budget and resources allocated to technology projects and programs. Therefore, it is important to manage costs for chatbot development and operations. Overspending on a chatbot that doesn't meet performance expectations or neglecting cost-optimization efforts can strain budgets and limit the resources available for other essential public services. In Chapter 9, we will dive deeper into various methods to optimize costs.

6.5 Example Use Cases

We covered the high-level architecture and various considerations for using GenAI-powered chatbots in the public sector, now it's time to consider example use cases. GenAI has widely increased the potential uses of chatbots, and there are hundreds of potential use cases. As we discussed earlier, we can generally classify the use cases into two categories:

1. Internal-facing (internal to the PSO)

2. External-facing (for constituents)

In this section, we discuss an example use case in each category in detail. We also include a use case for knowledge search; although knowledge search is not a chatbot, the architecture and applications are very similar to those of chatbots. Other use cases in general follow the same patterns and can be implemented in similar ways in alignment with the high-level architecture.

6.5.1 Use Case 1: HR Self-Service

Problem

Hannah, the head of the HR department in a PSO, faces many repetitive questions from employees on a daily basis about the organization's benefits, personal compensation, time-off policies, and other common topics. This takes up a significant amount of time and effort for her staff to provide reference policy documents and answer the same questions again and again. In addition, employees ask her staff to complete routine tasks such as generating employment verification letters, updating a home address or benefit selection, and submitting time-off requests, all of which are manual, repetitive work for humans.

Solution

The CIO offers to develop a GenAI chatbot to address Hannah's problem. She sets up a cross-functional team to implement the project. The team implements the high-level architecture in Figure 6-1.

Developers design a front end as a pop-up chatbot window on the HR department's website. They also enable the same chatbot window via employees' personal pages. They develop agents that connect to databases that store employee information. These agents retrieve relevant employee data and perform routine tasks such as employment verification. The developers implement content filtering to ensure that only appropriate information is presented in the chatbot.

Data engineers develop processes to upload documents from the HR system to the repository. Information from these documents is embedded into a vector database and passed as context to the FM using the RAG approach.

Data scientists evaluate various FMs and select the most appropriate FM using the guidelines provided in Chapter 3.

Prompt engineers develop prompts and test the chatbot extensively to ensure it answers relevant questions. They include several prompts that ensure that incorrect information is not passed back to the user. These prompts are saved by the orchestrator and are added to the queries made by the user.

Outcome and Benefits

The HR chatbot helps streamline Hannah's department operations while also improving the experience of employees. The application can provide instant and accurate responses for common inquiries as well as execute simple tasks. This largely reduces the need for employees to wait for staff to respond. It also allows employees to get assistance even after working hours. On the other hand, since most of the simple and repetitive tasks are handled by the application, Hannah's staff can focus on more complex tasks that require human effort.

Overall, the GenAI-powered HR chatbot helps improve the PSO employee satisfaction with self-service capabilities and helps Hannah's team to be more productive

Additional Considerations

Let's discuss some additional considerations for this use case.

- **Information update**: Department policies may be refreshed and modified every few years. Developers need to ensure that the vector database is refreshed periodically. This ensures that the application provides employees with the most recent information. Depending on the frequency of changes, this refresh can be done manually or using automated mechanisms.

- **Data privacy and security**: Security specialists ensure that data privacy and security measures are implemented. Developers need to ensure that each user interaction session is separate and independent. They should also implement encryption and explore additional protection mechanisms such as two-factor authentication.

- **Bring in human service**: While chatbots excel at handling many routine tasks through question and answer interactions, there will be situations requiring human intervention for more complex issues. Therefore, determining the most effective way to bring in human support is important. Developers should include a feature that provides the user with an HR staff's email address or phone number when the chatbot is unable to answer a question. Alternatively, they can design a dedicated button on the UI. When an employee clicks this button, a human resources staff member joins the chat to address the issue.

This use case demonstrates how a PSO can effectively leverage the power of GenAI to improve internal operations. Let's now turn our attention to another use case for improving internal operations. In this case, however, we discuss using an internal GenAI-powered search engine which follows a similar architecture to that of the chatbot.

6.5.2 Use Case 2: Internal Knowledge Search

Problem

A PSO in public health has a huge repository of internal data and documents related to topics such as disease control and prevention guidelines, vaccination programs, epidemiology reports, and budgetary allocations. Funding is allocated to different public health initiatives and programs. John, a caseworker, needs access to this information for his daily interactions with constituents. He typically relies on keyword searches within a document management system to find this information. However, many of his searches provide irrelevant results, requiring him to spend a lot of time browsing through numerous documents.

Solution

Mary, the IT director, decides to replace the current search capability with a search engine using GenAI. Her team implements the high-level architecture in Figure 6-1 to set up this search engine. Instead of developing a UI for a chatbot, developers design a UI that is similar to a Google search page. However, it contains a drop-down menu that John can use to select different topics of interest. These topics are passed to the orchestrator. The developers implement content filtering to ensure that only appropriate information is presented in the UI.

Data engineers develop processes to upload documents from the document management system to the repository. Information from these documents is embedded into a vector database and passed as context to the FM using the RAG approach.

Data scientists evaluate and use multiple FMs to handle different types of topics using guidelines provided in Chapter 3. They also determine which FM needs fine-tuning based on the vocabulary in each topic. ML engineers configure the orchestrator to route queries to the best FM suitable for the topic selected. They fine-tune the models selected by data scientists as needed.

Prompt engineers extensively test the search engine using different questions. They tailor default prompts in the orchestrator to provide the best results. These defaults are added to the questions posed by John in the search engine.

Outcome and Benefits

The new GenAI-powered internal search engine dramatically changes John's experience. He finds that he is able to ask questions that generate comprehensive responses including links to documents that he needs. This helps him especially when he needs to quickly locate relevant data and documents during disease outbreaks, program implementation, or responding to public inquiries.

Additional Considerations

The team needs to take into account the following additional considerations:

- **Data screening and classification**: Data engineers should put rules in place to remove or redact sensitive information during data migration to the repository.

- **Information update**: Since data in these systems may change frequently, engineers need to consider frequent updates of the repository using automated jobs.

- **Data access**: Not all data and documents at a PSO are available to every employee. For example, John should not have access to HR information. To accomplish this, developers set up permissions and policies which restrict John's ability to view certain topics. They can also implement fine-grained access control to restrict access at a more detailed element level.

6.5.3 Use Case 3: Constituent Chatbot

Problem

A PSO involved in social services often faces high call volumes and long wait times for constituents seeking information or assistance with various programs. Jane is one such constituent. She is a single mother and needs to frequently interact with the PSO for several things. For example, she needs to determine eligibility for benefits such as food stamps, housing assistance, and childcare subsidies. She also needs to track the status of applications and benefits. She gets frustrated when she is unable to get answers to her questions quickly. At the same time, John, the caseworker in the PSO, is under constant pressure. He has a number of complex cases to work on, but a lot of his time is spent on routine inquiries and calls.

Solution

The CIO decides to implement a public-facing chatbot on their website to address these challenges and forms a cross-functional team. The team implements the high-level architecture in Figure 6-1. The architect works with the team to review and incorporate techniques for the performance and scalability of the application as will be discussed in Chapter 9.

Developers design a user-friendly chatbot interface accessible on the PSO's website. They configure an agent to connect to relevant PSO databases to retrieve program information and application status and initiate actions such as address changes. They also set up content filtering guardrails that ensure that any harmful or inappropriate content is prevented from either being sent to the FM or displayed to Jane.

Data engineers develop processes to upload documents needed to the repository. Information from these documents is embedded into a vector database and passed as context to the FM using the RAG approach.

Data scientists evaluate the FMs needed using guidelines provided in Chapter 3. They also determine the type of FM (LLM, MLLM) or FMs and if the models need fine-tuning based on the vocabulary.

Prompt engineers develop prompts and test the chatbot extensively to ensure it answers relevant questions and does not have performance issues. They include several prompts that ensure that incorrect information is not provided. Because this is an external-facing application, prompt engineers review the outputs with policy analysts to ensure that no policies or regulations are violated. The prompts are saved by the orchestrator and are added to the queries made by the users such as Jane.

Outcomes and Benefits

Jane is extremely pleased with the new chatbot. Not only does it offer 24/7 availability, but it also provides answers to her question quickly with a number of self-service capabilities. John finds that he has a lot more time to work on complex cases and help additional constituents with advanced needs.

Additional Considerations

The team needs to take into account the following additional considerations:

- **Continuous evaluation and audits**: Since this chatbot is external facing, independent auditors should periodically evaluate the output of the models. They can use metrics discussed in Chapter 4 for the evaluation. The auditors should be trained on using the chatbot and understand how to flag issues or inconsistencies.

- **Information update**: Similar to those in the HR example, data engineers may need to refresh the repository frequently to provide the latest program details, eligibility criteria, and policies.

- **Data privacy and security**: Security specialists need to make sure robust security measures are in place to safeguard constituent data collected through the chatbot. This includes encryption, secure data storage, and access control mechanisms.

- **Human intervention**: Developers should incorporate logic to seamlessly transfer complex inquiries or issues requiring human judgment to the PSO staff. This can be achieved through features such as offering contact information for specific caseworkers or initiating live chat sessions with agents.

- **Continuous monitoring**: DevOps engineers need to incorporate continuous monitoring of the application to identify and resolve issues.

By implementing a public-facing chatbot, PSOs can significantly enhance service delivery, improve staff productivity, and improve overall constituent experience.

6.6 Conclusion

This chapter explored how GenAI and the RAG architectures empower the development of powerful chatbots for PSOs. We highlighted the key differences between GenAI-powered chatbots and traditional rule-based chatbots, emphasizing the superior capabilities of GenAI in understanding natural language, providing informative responses, and adapting to diverse user queries.

We discussed a high-level architecture for these chatbots, outlining the key technical considerations involved. We then explored illustrative use cases, showcasing how both internal-facing and external-facing chatbots can be implemented to significantly enhance public service delivery. We also covered how the chatbot architecture can be extended to implement a GenAI-powered search engine.

While this chapter focused on specific use cases, it is important to recognize the vast potential of this architecture for diverse applications. Different front-end formats, data ingestion methods, and agent tasks can be tailored to address various needs within a PSO.

In essence, GenAI-powered chatbots empower PSOs to offer more responsive digital public services while simultaneously unlocking operational efficiencies. By carefully considering the nuances involved in implementation, these solutions can fundamentally transform how governments interact with both employees and constituents.

In the next chapter, we will dive deep into how GenAI can be utilized to summarize lengthy documents, reports, research papers, public comments, and textbooks. By leveraging GenAI for document summarization, PSOs can unlock significant potential for time savings, resource optimization, and cost reduction.

CHAPTER 7

Summarization

Public sector organizations (PSOs) often have to navigate through complex and diverse information. In the realm of the public sector, information overload is a constant struggle. Government agencies, research institutions, and public organizations are inundated with massive volumes of data, reports, documents, and communications on a daily basis. As an example, the US public sector is potentially the largest producer of data.[1]

Let's take the example of Xiulan, the administrator of a PSO. Xiulan frequently needs to review information related to the PSO. However, she finds that manually summarizing the vast amount of information available is not only time-consuming but also resource-intensive. Much of the information contains domain-specific terminology, legal details, and other intricacies. Additionally, she receives information from various sources, such as internal reports, external stakeholders, public comments, and social media.

GenAI offers powerful options to address such challenges. By leveraging Foundation Models (FMs), Xiulan can benefit from automated summarization capabilities, enabling her to generate concise and coherent summaries of lengthy documents, reports, or other text-based information. This not only reduces the time and effort required for manual summarization but also ensures consistent and timely access to crucial information.

---

[1] https://digital.gov/2018/03/14/data-briefing-value-federal-government-data/

© Sanjeev Pulapaka, Srinath Godavarthi and Dr. Sherry Ding 2024
S. Pulapaka et al., *Empowering the Public Sector with Generative AI*,
https://doi.org/10.1007/979-8-8688-0473-1_7

In this chapter, we examine the high-level architecture and considerations related to developing a custom GenAI application for summarization. We then discuss multiple use cases relevant to the public sector.

7.1 High-Level Architecture

In the previous two chapters, we discussed use cases where the tasks completed by an FM, such as a Large Language Model (LLM), Multimodal Large Language Model (MLLM), and text to image, are more interactive in nature. For example, in the case of generating a document or an image, the user enters a prompt and typically waits for a response from the model. In the case of chatbots, the user types questions and receives answers.

In the case of summarization, there are two possibilities we need to account for in the architecture. We could have a user actively interacting with the application as seen before. In this case, they may upload a document and define a prompt that instructs the FM to summarize the document. However, we could also have a process where an FM summarizes text without an active user involved. In this case, the orchestrator passes the summarization prompts to the FM. Figure 7-1 illustrates the high-level architecture of an application that incorporates both possibilities.

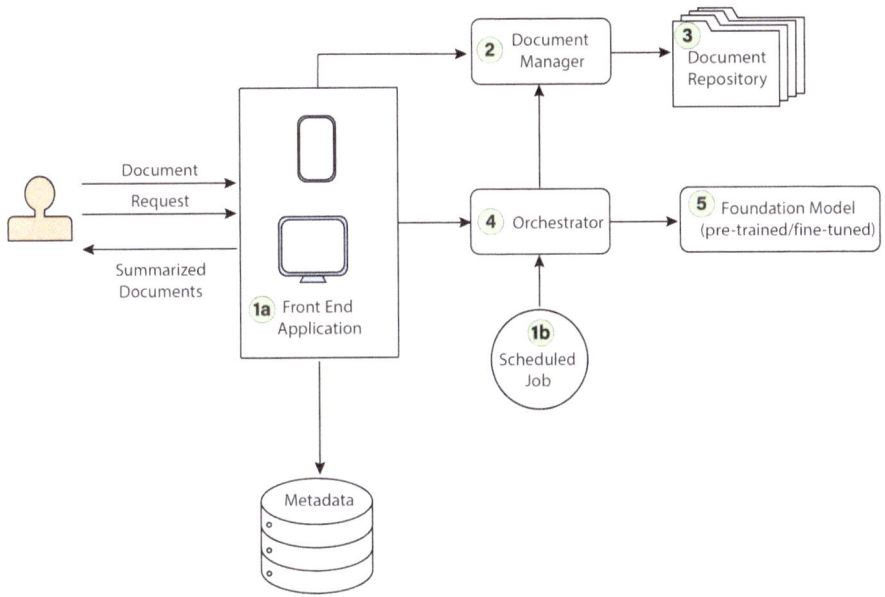

Figure 7-1. *Document Summarization Architecture*

This architecture is a slight modification of the high-level architecture illustrated in Chapter 4, Figure 4-4. The main difference is the addition of (1b) Scheduled Job. Let's walk through each component again, examining the relevance to the summarization use case.

1. **1a. Front-end application**: The user interacts with a front-end application that presents an interactive user interface (UI). The user is any person seeking to view or review document summaries generated by the FM. The user could also upload documents and request summaries of that document. Users acting as reviewers should be able to edit the document to make any modifications necessary. You can have manual reviews or automated reviews by sending the generated content back to another FM to review the accuracy of the content. Therefore, the UI

should have the ability to flag sections of content which are not accurate or satisfactory. These flags should be stored in a metadata repository so that you can use the information for further tuning of prompts.

1b. Scheduled job: This component is a trigger that kicks off a program managed by the orchestrator. The program contains a prompt with instructions to summarize the document as well as the code to execute that prompt against the model. This is helpful when you need to process large documents or a number of small documents regularly. In these situations, the user does not need to wait for the processing to be complete because the entire summarization is implemented by the scheduled job.

2. **Document manager**: The front end interfaces with a document manager that is responsible for storage, organization, search, and retrieval of documents in a repository.

3. **Repository**: The repository stores source documents that need to be summarized as well as summarized documents generated by the FM. This typically includes unstructured documents in various formats such as PDF, spreadsheet, and so on.

4. **Orchestrator**: The front-end application invokes an orchestrator that does the heavy lifting by interacting with the FM for the content summarization. The orchestrator defines specific prompts that guide the FM to summarize a document. There are

different summarization strategies depending on the length of the document. We'll discuss these in detail in the next section.

5. **FM**: This can be an FM of choice, either the base model or the fine-tuned model. As mentioned earlier, fine-tuning may be required if you have very specific domain nomenclature or very specific summarization-related tasks.

7.2 Key Considerations

Let's review some key considerations for summarization architecture:

7.2.1 Batch or Real-Time Processing

Any given PSO may have two types of requirements for summarization: batch and real time. In batch processing, you process summarizations at predefined frequencies. In the preceding architecture, this processing is implemented using the scheduler. High volumes of predictable documents, such as monthly reports, are well suited for batch processing.

Conversely, in real-time processing, the user requests summaries on demand and expects a quick response. Real-time processing offers immediate access to summaries, improving user experience for constituent portals or internal inquiries.

Real-time processing use cases need thorough testing and evaluation. There is no time to have the response reviewed before it is presented to the user. Batch processing on the other hand has a significant advantage since, once processing is complete, you can have human reviews to ensure that the responses are accurate.

In some situations, a hybrid approach combining both batch and real-time processing can be the best solution. The key is to continuously evaluate the effectiveness of your approach based on user feedback and evolving needs within your PSO. This allows you to leverage summarization effectively, striking a balance between accuracy and responsiveness to constituent needs.

7.2.2 Context Window Size and Length of Document

The *context window* of an FM (most likely LLM or MLLM for summarization) is the amount of information it can process at once. This plays a crucial role in how well the model can summarize the report.

- **Large context windows**: Some models, such as Anthropic's Claude v2 and v3, have a large context window (200k tokens which is roughly equivalent to around 350 pages of A4 text). This allows the models to effectively process and summarize even lengthy reports such as city budgets or complex policy documents.

- **Limited context windows**: Other models have smaller context windows. If you are using such a model, you might need to break down the report into smaller sections to fit within the window, for example, summarizing a long public health report in chunks based on specific sections such as disease trends or vaccination rates.

Strategies for Summarizing Long Documents

Imagine you have a lengthy government report summarizing the findings of a city-wide survey. Let's review some strategies to tackle summarizing this report:

- **Summarize by sections**: Divide the report into chapters or subsections. Use the FM to summarize each section, then use the same or another FM to create a summary of these summaries. This is similar to having a team of analysts provide summaries of different parts of a complex policy document, followed by a final high-level overview.

- **Summarize by topics**: First, identify key topics using manual methods or by using machine learning algorithms such as Latent Dirichlet Allocation (LDA) and nonnegative matrix factorization (NMF). This is similar to analyzing the report for topics such as infrastructure, education, or public safety. Then, split the report into chunks and provide clear prompts to the FM to summarize each chunk based on those topics.

- **Hierarchical summarization**: Break down the report into fixed-size chunks and generate summaries for each. Then, use the FM to create a higher-level, concise summary of the entire report. This is similar to having a hierarchical report structure where each section has its own summary, followed by a final executive summary.

- **Query-driven summarization**: Leverage the question-answering capabilities of FMs. Provide the report and specific questions about its content to the FM. The FM can then generate a summary that focuses on those questions, effectively highlighting relevant parts of the report.

Using these strategies, PSOs can effectively summarize large documents such as reports, transcripts, or regulations, making the documents more accessible and digestible for constituents and internal stakeholders.

197

7.2.3 Cost Control

Summarizing long documents can lead to increased costs. Costs can vary depending on the type of deployment (self-hosted or cloud based). Cloud providers offer different types of pricing models. Some pricing models are based on the number of tokens processed; summarizing large documents will result in high costs, not only in production but also in development and testing. There are two approaches to mitigate this. First, PSOs can consider exploring other types of pricing models available, such as time-based charges. Second, prompt engineers should optimize and evaluate prompts used by the orchestrator such that the number of round trips or back and forth exchanges between the orchestrator and FM are reduced.

So far, we have discussed the high-level architecture and key considerations for summarization. Now, let's dive into example use cases.

7.3 Example Use Cases

7.3.1 Summarization Use Case 1: Policies and Guidelines

Problem

In the public sector, policies and guidelines often span numerous documents containing extensive information and legal details. John, a caseworker in one such PSO, finds it challenging to comprehend and adhere to these policies and guidelines effectively. John is continuously evaluating new policies and helping the constituents such as Jane with eligibility rules for benefits. This results in lengthy wait times for the constituents in getting responses.

Solution

The CIO decides to use GenAI for summarizing large and complex policy documents. Her team implements the high-level architecture illustrated in Figure 7-1.

However, instead of creating a separate front-end application, they simply integrate the repository with the PSO's content management system. However, they face a challenge: policies and guidelines frequently contain links to other documents and websites. Data engineers perform additional data processing to resolve this issue. They create a program that searches for hyperlinks in the documents and downloads relevant text to the repository.

In a similar vein, many policy documents may contain images. Data engineers use one of the techniques mentioned in Chapter 6 called grounding modalities into a primary modality (discussed in Section 6.4.3) to solve this problem. They separate out images from each document and use an MLLM to generate textual descriptions and metadata for each image. This metadata is added back to the original document alongside the image.

Data scientists evaluate and select the best MLLM for the use case using guidelines provided in Chapter 3.

ML engineers evaluate some of the summarization techniques mentioned in the previous section to determine the best approach for policies. They implement this approach in the orchestrator, making sure that costs are kept under control. Prompt engineers develop and test the prompts necessary to create summaries of the documents.

Outcome and Benefits

John is very pleased. He can now find concise summaries of each policy and guideline he is interested in. He is able to understand the summaries quickly, saving him a lot of time and frustration. As a result, he can get through a lot more of his cases quickly, improving the overall efficiency of case processing in the PSO. He can also help the constituents with their queries, thus improving their overall experience.

Additional Considerations

- **Model training and customization**: Data scientists need to evaluate if the FMs require fine-tuning for domain-specific data.

- **Human oversight and review**: Developers need to incorporate workflows to have subject matter experts review and validate the summaries to ensure accuracy, completeness, and compliance with legal and regulatory requirements.

- **Data privacy and security**: Security specialists need to verify that developers have incorporated appropriate measures to ensure data privacy and security, such as using secure data storage and transmission protocols and implementing access controls. We will discuss privacy and security in more detail in Chapter 9.

7.3.2 Summarization Use Case 2: Research Paper Abstracts

Problem

Many PSOs are involved in different types of research on various topics ranging from science, economics, law, politics, social sciences, and socioeconomics. Wei is a researcher in one such PSO. Wei needs to create abstracts as part of his research publications. These abstracts are present in various catalogs and are one of the best ways for researchers such as

Wei to share and find relevant information. Wei spends significant time in creating effective summaries of research papers for the purpose of abstracts. He would rather spend that time in doing more research which would ultimately benefit constituents and society as a whole.

Solution

Mary, the IT director, decides to address Wei's issue by automating abstract creation using summarization with GenAI. She establishes a team to implement the architecture described in Figure 7-1.

Research papers rely heavily on citations; data engineers mine documents to find and import citations to the repository.

Data scientists review and determine that fine-tuning the FM is essential for a number of reasons. For one, the research involves very specific terminology; this is called domain-based fine-tuning. Second, research papers often have specific sections with unique formats such as introductions, methods, results, discussions, and conclusions. Fine-tuning the FM using task-based fine-tuning can help identify these sections and prioritize information from specific sections. Third, task-based fine-tuning is necessary when using labeled datasets of research papers and their corresponding human-written abstracts. Last, but not least, abstracts typically have strict word count limitations. Task-based fine-tuning the FM is necessary to generate concise summaries that capture the essential information within the specified word limit.

ML engineers fine-tune the FMs based on data scientist recommendations. They configure the orchestrator as needed depending on the type of method used for summarization as discussed earlier. They also ensure that the orchestrator is optimized to keep costs under control.

Outcome and Benefits

Wei is thrilled that GenAI can analyze his full research paper and generate a concise summary, saving him significant time and effort compared to the time he spends in manual writing. He also realizes that the summaries are concise, clear, and easier to read.

Additional Considerations

- **Novelty detection**: Research often builds on previous work. Richard may need to fine-tune the FM to identify the novel contribution of the research papers summarized.

- **Focus and bias**: Researchers might be interested in specific aspects of the paper (e.g., methodological advancements). Therefore, Wei should ensure that the summaries capture these aspects. He needs to ensure that potential bias introduced during fine-tuning does not favor certain aspects of the paper over others. During fine-tuning, ML engineers should work actively with Wei to ensure a balanced and accurate summary.

7.3.3 Summarization Use Case 3: Comment Summarization

Problem

Public comments can be voluminous and unstructured, containing a wide range of opinions, concerns, and suggestions. Xiulan, the administrator of a PSO, is responsible for reviewing comments on rules and regulations. The review is required to ensure that the PSO takes into account public perception of impending rules. However, Xiulan's team spends a significant amount of time reviewing the comments. Xiulan wishes she could get summaries of the comments so that her team can quickly get through the reviews.

Solution

Mary, the IT director, determines that GenAI can be used to effectively summarize comments. Her team implements the architecture in Figure 7-1. It is very important to ensure that Personally Identifiable Information (PII) is not revealed in any manner. Therefore, data engineers scrub all comments to redact PII and load the scrubbed comments to the repository. Prompt engineers realize that in addition to simply summarizing the comments, it may also be useful to identify the sentiment (positive, negative, neutral) expressed in each comment. Accordingly, they define prompts in the orchestrator that instruct the FM to identify the sentiment of the comment in addition to summarization. They extensively test the prompts by having the summaries and sentiments reviewed by Xiulan's team members.

Outcome and Benefits

Xiulan's team is able to process comments quickly by reviewing summaries of comments as opposed to reading through lengthy comments. In addition, sentiment analysis allows them to get an idea on the opinion of the public. For example, they can aggregate sentiments by topic or category and understand the overall opinion.

Additional Considerations

- **Mitigating bias**: Public comments might be skewed toward certain demographics or viewpoints. Therefore, comments need to be periodically reviewed by auditors to ensure that biased opinions are not being generated.

7.4 Conclusion

In this chapter, we saw how summarization using GenAI offers immense potential for PSOs. By leveraging the power of FMs, PSOs can automate the process of summarizing lengthy documents, reports, research papers, and public comments, among others. This not only saves time and resources but also ensures that critical information is readily accessible and digestible. The key to successful implementation lies in various factors such as batch vs. real-time processing requirements, the context window size and length of the document, and the specific nuances of the summarization task at hand. By adopting techniques such as hierarchical summarization, query-driven summarization, and fine-tuning models for domain-specific vocabularies and summarization structures, PSOs can tailor the summarization process to their unique needs.

As with any AI-driven solution, it is crucial to maintain human oversight and review to ensure accuracy, fairness, and ethical compliance. PSOs should also prioritize data privacy and security, particularly when handling sensitive information. With the right approach and safeguards, GenAI-powered summarization can positively impact how public sector organizations process and disseminate information, improving internal operations and efficiencies.

The next chapter will delve into another critical aspect of public sector operations: program management, business intelligence, and reporting.

These functions are essential for tracking progress, analyzing data, and making informed decisions that drive organizational success. By harnessing the power of GenAI, PSOs can streamline and enhance these processes, unlocking new insights and optimizing resource allocation.

CHAPTER 8

Program Management, Business Intelligence, and Reporting

Public sector organizations (PSOs) are subject to a myriad of federal, state, and local regulations, necessitating effective program management and compliance reporting. They must provide transparency in their operations to maintain public trust. Providing a high level of transparency requires accurate reporting of program metrics and key performance indicators in alignment with mission outcomes. Failure to comply with reporting requirements can result in the loss of funding, underscoring the importance of robust reporting capabilities. As an example, the US Congress authorized the Center for Medicare and Medicaid Services (CMS) to withhold funds from states that do not satisfy certain reporting requirements.[1]

---

[1] www.cbpp.org/research/health/unwinding-watch-tracking-medicaid-coverage-as-pandemic-protections-end?item=28546

© Sanjeev Pulapaka, Srinath Godavarthi and Dr. Sherry Ding 2024
S. Pulapaka et al., *Empowering the Public Sector with Generative AI*,
https://doi.org/10.1007/979-8-8688-0473-1_8

PSO leadership and employees face numerous challenges in the areas of program management, reporting, and analytics. Maria, the Chief Information Officer (CIO), is primarily responsible for managing regulatory risk and compliance for her PSO. Xiulan, the administrator, is tasked with ensuring that her PSO operates in a transparent manner by publishing key program performance metrics to the government and public. Xiulan often struggles with data-driven decision-making because she doesn't have the access to capabilities that can provide her deeper mission insights and intelligence.

Both Maria's and Xiulan's staff are often burdened with arduous tasks related to program management, analytics, and decision-making, which can be time-consuming and resource-intensive. Putting together high-quality presentation decks and compiling reports for executives and dashboards for federal compliance can take hours, diverting valuable time and resources away from core operations. Both these executives are heavily dependent on Mary, the IT director, to provide them with capabilities that can help with program compliance and reporting.

However, Mary's team often struggles with existing and legacy reporting systems that present several challenges. Traditional reporting systems are often highly complex and inflexible, offering static reports without interactive capabilities.

The preceding examples are some of the challenges faced by PSOs in today's dynamic business environment. Leaders and managers require interactive exploration, data analysis, and visualization to make informed decisions. Traditional reporting systems lack advanced features that leverage AI capabilities for predictive modeling and "what-if" scenario analysis. Last but not least, program leaders often have to depend on technology and development teams for even minor changes to reports, such as formatting adjustments or customized views. The turnaround time for such requests can be unacceptably long, hindering timely decision-making.

To address data-driven decision-making, the US federal government has mobilized a unified effort to undertake one of the largest federal transformation initiatives.[2]

Generative AI (GenAI) offers promising capabilities and solutions that can address these challenges. In this chapter, we examine these capabilities.

8.1 Broad Areas Where GenAI Can Assist with Reporting, Business Intelligence, and Analytics

First, we explore the various ways in which GenAI can help PSOs address the preceding challenges. We will dive deeper into specific use cases later in this chapter.

8.1.1 Report Generation

PSO employees can generate reports using simple natural language prompts. Traditionally, this was implemented using Structured Query Language (SQL), a programming language for storing and processing information in a relational database. However, that approach required acquiring SQL tools and skills. Now, that is no longer needed with GenAI – you can simply type in your question, called a natural language query, as a prompt, and the Foundation Model (FM) will generate the results. As a result, business users can directly work with data, as opposed to relying on IT staff. IT staff can now focus on providing high-quality data and the systems to interface with the data including GenAI.

---

[2] www.forbes.com/sites/randybean/2022/06/01/how-the-us-federal-government-is-mobilizing-to-enable-data-driven-decision-making/

8.1.2 Business Intelligence

FMs can be leveraged to provide deep insights into program trends, budgets, mission outcomes, and objective key results (OKRs). Again, this can be done without programming and heavy lifting that was required previously.

8.1.3 Analysis of Large Datasets

FMs can directly analyze structured and unstructured data sources to detect anomalies, extract insights, and identify trends, which can inform policy, legislation, and decision-making processes. Imagine using this power of GenAI on the 300,000 or so datasets that are available on data. gov,[3] including geospatial data, census data, healthcare data, and so on. This can help with generating valuable insights into weather patterns, census trends, or disease trends within the country.

8.1.4 Data Visualization and Storytelling

FMs can visualize data and generate narratives from existing unstructured data, enabling users to better understand and communicate insights. This is done using both analytical as well as content generation capabilities.

8.1.5 Predictive Analytics

FMs can forecast budgets, enrollment numbers, and other metrics based on historical trends, aiding in demand planning for resources. Additionally, they can help predict public health and disease trends.

---

[3] https://data.gov/

8.1.6 Interactive Data Exploration and "What-If" Analysis

Similar to report generation, FMs enable users to explore and interact with data using natural language, enabling users to ask questions, request insights, explore different scenarios, and inspect visualizations in real time.

8.1.7 Document Analysis and Insights

FMs can extract insights from existing documents. For example, you can analyze earning reports of an organization, which are typically generated in a PDF format.

Furthermore, FMs can facilitate document comparisons, automating the process of comparing, analyzing, and summarizing information from different documents. This significantly increases the speed and accuracy of document comparison, enabling users to make informed decisions faster and with greater confidence.

Some of the capabilities discussed earlier can be obtained directly from existing products and solutions in the marketplace. For example, many vendors in the business intelligence space have started providing the ability for natural language–based queries in their products. However, some PSOs, especially those that have complex internal systems, may need to develop internal solutions using FMs.

In the next section, we introduce a high-level architecture for developing these capabilities.

8.2 Business Intelligence, Data Analytics, and Reporting High-Level Architecture

Figure 8-1 depicts a high-level architecture diagram for business intelligence, analytics, and report generation using FMs. The core components of this architecture include

- **Front-end application or user interfaces (UIs)**: This includes the functionality to accept user inputs, suggest query prompts, render, format, and paginate the reports. UI acts as the primary component for users to interact with the system.

- **Prompt library**: This component is primarily used to provide sample prompts or prompts that have successfully worked in the past.

- **Orchestrator**: This module provides capabilities for managing sequences of steps for interacting with the FM.

- **Metadata catalog**: This provides the catalog for table names and a list of data sources.

- **Data sources**: Sources of data for analytics and reporting.

- **FM**: The FM that provides the functionality to convert natural language queries to SQL queries as well as convert the results obtained back from the query to natural language. This could be an LLM or an MLLM.

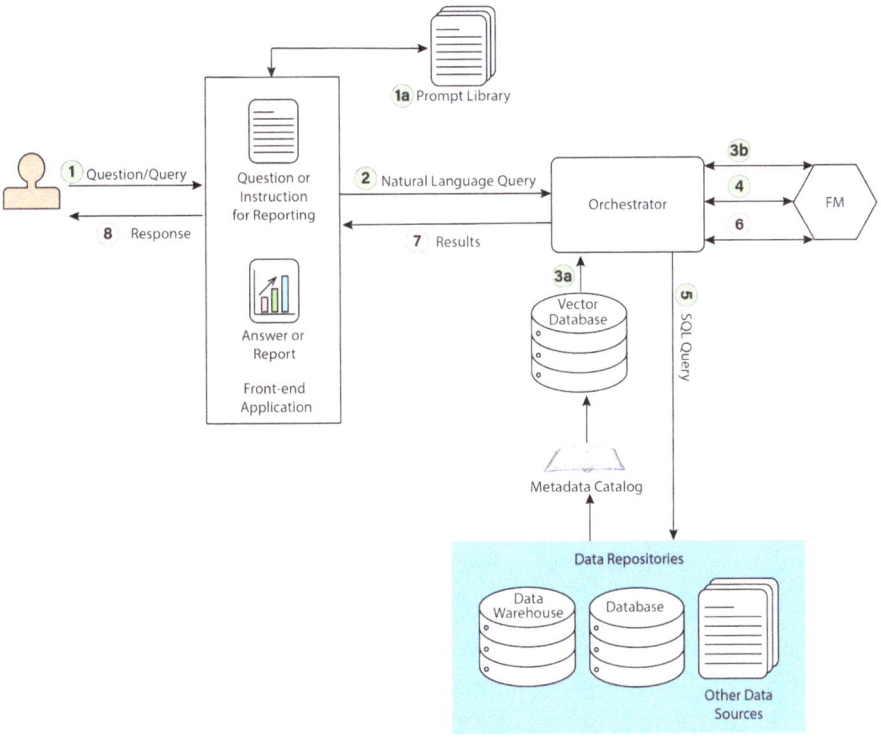

Figure 8-1. *Business Intelligence, Analytics, and Reporting Architecture*

8.2.1 Workflow Steps

The diagram in Figure 8-1 also includes the steps a user takes when they interact with an application that uses the architecture.

Steps 1, 1a: The user interacts with the front-end application via a user interface (UI). The UI has the capabilities to provide report generation prompts and prompt engineering suggestions using a prompt template library. For example, if the user needs to get the total unemployment insurance claims during the pandemic sorted by each state, they can type in the following prompt:

```
Display the total unemployment insurance
claims during the pandemic for each state
sorted in descending order. Format the
output with a bar graph
```

This type of prompt is also known as a Natural Language Query (NLQ). In other words, it is a question asked in natural language as opposed to the traditional query using SQL as described earlier.

Step 2: The front-end application submits the NLQ to the orchestrator.

Step 3a: This step applies Retrieval-Augmented Generation (RAG), discussed in detail in Chapter 4, to generate table names from the metadata catalog. Applying the RAG workflow, the NLQ is first compared to table and column names in the vector database, and the most closely matching values are returned. Note that we don't show an embedding model to keep the diagram simple. However, it is needed to create the vectors from the metadata as well as the NLQ; refer back to Chapter 4 for specific details on this approach.

Step 3b: The NLQ and the potential choices for table and column names are sent to the FM to get the exact table and column names. For example, for the NLQ earlier, the FM determines that the table name is "Unemployment Insurance Claims" and the column "States Unemployment Insurance."

Step 4: This step generates a SQL query. In this case, the NLQ and the corresponding table and column values from step 3b are sent to the FM. The FM transforms these into an SQL query and sends it back to the orchestrator.

Step 5: In this step, the orchestrator runs the query on the database to retrieve the results.

Step 6: The result from the SQL query is passed to the FM. The FM converts the result into natural language.

Step 7: The natural language response is passed back to the UI where it is further formatted for end-user consumption.

Step 8: The user receives the final report content on the UI.

8.3 Key Considerations

8.3.1 Front-End Application and UI Design

The front-end application has significant functionality embedded into it including providing the right prompts to the user, sending the prompt/query content to the orchestrator, and rendering the results. Therefore, to enhance the user experience, the UI should be intuitive to use. The UI should include user feedback mechanisms for each prompt; for example, using a simple thumbs up/down icon to indicate success or failure. This feedback can be used to track and improve the effectiveness of prompts. Also, the front-end application should run seamlessly across multiple devices (such as on laptop, mobile phone, tablet, and so on). Last but not least, the application should scale to handle a large number of users with low latency and fast response times including the content display.

8.3.2 Prompt Design and Prompt Template Library

Prompt design is important in producing the desired output, providing optimal user experience, and in minimizing repeated prompts (and traffic) back to the application. Specifically for data analytics and reporting, avoid query failures; this will help reduce the load on databases, infrastructure, and services and will help reduce costs. Review guidance provided by the application to develop efficient prompts to get accurate results. This is where a prompt library that contains validated prompts is very helpful.

Also, as discussed earlier, you can use the concept of iterative learning where you continuously refine prompts based on the insights gained from user experience and feedback.

8.3.3 Query Handling, Integration with Multiple Metadata, and Data Sources

NLQs can become complex very quickly. The user has no knowledge of whether or not a specific NLQ leads to a full table scan. A full table scan implies that all the rows of a table are processed by the database query. This can be very time-consuming. As an example, users can submit a simple prompt, such as "Give me the census data." This prompt can result in a SQL query that scans the entire census data table from the inception date for 300+ million people in the United States: this could take days to complete.

The preceding process can be further compounded if thousands of users are submitting similar prompts and NLQs. Therefore, carefully design the application to prevent such wide-open queries and minimize the risk of resource exhaustion on the back-end data sources and infrastructure. Close monitoring of queries, including the time taken and the type of SQL queries that each NLQ is generating, would help providing specific guidance to the user. As an example, for the preceding query, the system could respond by saying

```
Sorry, please provide specific timeframe
and/or other filters with your query -
example: you can say - Provide me the census
data for 2020 with the total population in
California broken down by zip code
```

Ultimately, this becomes an exercise of helping the user to become more specific about what they need.

Additionally, the NLQ may need data from multiple sources; techniques such as data aggregation may be needed to improve performance of the application. Other factors such as how frequently the metadata is updated and which tools to use for the same also become important considerations.

8.3.4 Cost

Finally, it is essential for you to understand the cost implications of using an FM for reporting and analytics. Large datasets can increase cost and latency. We will discuss this in more detail in Chapter 9.

Now that we have discussed the architecture of a custom application for reporting and analytics, let's dive deep into some of the use cases. Some of these use cases may apply this architecture, while others may directly use available products on the market.

8.4 Data Analytics and Reporting: Use Cases

8.4.1 Use Case 1: Compliance and Reporting

Problem

There are a broad set of compliance and reporting requirements across a number of programs in the public sector; as an example, the Government Accountability Office (GAO) in the United States issues guidance for federal agencies on financial management, auditing, and reporting.[4] This guidance includes specific reporting, such as project and expenditure reporting and recovery plan performance, among others.

---

[4] www.gao.gov/products/gao-22-105894

As another example, the Australian Accounting Standards Board (AASB) adopts International Financial Reporting Standards (IFRS) with some modifications. Xiulan is the administrator of a PSO in a healthcare organization. She is responsible for creating reports as part of these reporting guidelines. The business climate of healthcare organizations changes frequently due to onset of illnesses. Xiulan finds it difficult to adjust the reports quickly. The IT staff in the PSO is extremely busy, and changes to reports take time. However, Xiulan knows that failure to submit reports in time can result in reduced funding and/or penalties. Xiulan wishes there was a way she could generate the reports using self-service capabilities.

Solution

Maria, the CIO, determines that implementing a GenAI-based solution can address Xiulan's concerns. She puts together a team that implements the high-level architecture in Figure 8-1.

Developers create a front-end application that Xiulan can use to create the reports she needs. They also identify all the data sources needed for Xiulan and configure a metadata catalog. ML engineers create a vector database from the source of data identified and set up the workflow specified in Figure 8-1 using the orchestrator. ML engineers also configure the prompts in the orchestrator to convert natural language queries from Xiulan into SQL using the metadata catalog.

Data engineers work with ML engineers to ensure that the SQL being generated is optimized. They also work with database administrators to minimize latency of SQL using techniques such as materialized views, partitioning, and sharding.

Data scientists evaluate various FMs and select the most appropriate FM suited for reporting using the guidelines provided in Chapter 3. They also evaluate if fine-tuning is needed for the models.

Prompt engineers develop sample questions and test the front-end application extensively to ensure it answers relevant questions. They include several prompts that ensure that incorrect information is not passed back to the user. These prompts are saved by the orchestrator and are added to the queries made by the user.

Outcome and Benefits

Xiulan is very happy with the new system. She can now use simple natural language prompts to generate and format reports. For example, she is able to ask questions such as

```
What contributed to the decrease in
vaccinations for the month of November?
```

This dramatically improves her ability to meet reporting deadlines and reduce the risk of losing funding. It also helps IT staff by allowing them to focus on more complex activities.

Additional Considerations

Developers can expand the front end by introducing the capability to tell stories using data. They can do this by stitching together the results of the individual natural language queries by Xiulan. For example, Xiulan enters the following prompt in the front end:

```
Build me a story with vaccination trends
across the country with state by state
breakdown; provide me insights into why
the numbers have dropped for this year and
strategies on improving participation.
```

When this is entered, she receives an end-to-end story containing key points and data; between the request and response, internally

- The question is passed to the orchestrator.

- The orchestrator executes a prompt that asks the FM to break down this question into multiple individual questions.

- The orchestrator sends each individual question received back to the FM as a question.

- The orchestrator collects all the answers to each individual questions.

- The orchestrator sends the collected set of answers to the FM with the instruction to generate a story.

This capability helps Xiulan gather further insights and intelligence on critical program trends and program performance and develop measures accordingly.

8.4.2 Use Case 2: Improve Customer and Employee Experience

Problem

In the public sector, customer experience (CX)[5] is the public's perception of and overall satisfaction felt in communicating with any PSO. Customers could be a constituent, business, or an organization that interacts with a PSO.

---

[5] www.performance.gov/cx/

For an individual, good CX can mean less time in a government office, less time on the phone with a help desk, or more digestible information to support a decision. Good CX also means that the customers are getting high-quality services and that the PSOs are spending their resources on solving the right problems for constituents. Research[6] shows that US federal and state government services rank near or at the bottom in customer experience ratings, with most customers rating their experience as poor or very poor. As such, CX has been a major focus area for government agency leaders and PSOs. The US president has even issued an executive order, "Transforming Federal Customer Experience and Service Delivery to Rebuild Trust in Government,"[7] to emphasize this aspect and establish a number of commitments across the US federal government.

Marco is the Director of CX at a PSO responsible for managing unemployment benefits. He is tasked with improving the overall constituent experience while optimizing operational costs. Marco faces several challenges:

1. **Limited insights into service consumption patterns**: Marco lacks deep insights into how constituents interact with the PSO's services, making it difficult to identify areas for improvement and personalization.

2. **Reactive approach to constituent needs**: The PSO often responds reactively to constituent needs, such as during natural disasters or economic downturns, leading to delays and suboptimal resource allocation.

---

[6] www.mckinsey.com/industries/public-sector/our-insights/how-us-government-leaders-can-deliver-a-better-customer-experience
[7] www.whitehouse.gov/briefing-room/presidential-actions/2021/12/13/executive-order-on-transforming-federal-customer-experience-and-service-delivery-to-rebuild-trust-in-government/

3. **Inefficient contact center operations**: The PSO's contact center struggles with long call handling times, low agent productivity, and limited ability to analyze customer feedback and sentiment.

Marco is aware that failure to address these challenges could result in poor constituent satisfaction, inefficient resource utilization, and erosion of public trust in the PSO's services.

Solution

Recognizing the potential of GenAI to address these challenges, Marco consults with the PSO's CIO. The CIO proposes a GenAI-based solution in alignment with the architecture outlined in Figure 8-1.

The development team creates a front-end application that allows Marco and his team to interact with the GenAI system using natural language queries. They also identify and configure the necessary data sources, including citizen interaction data, service delivery metrics, and feedback surveys.

ML engineers set up a vector database from the identified data sources and configure the workflow specified in Figure 8-1 using the orchestrator. They also configure prompts in the orchestrator to convert natural language queries into SQL queries using the metadata catalog.

Data engineers work closely with ML engineers to optimize the generated SQL queries for performance and minimize latency using techniques such as materialized views, partitioning, and sharding.

Data scientists evaluate various Foundation Models (FMs) and select the most appropriate model for reporting and analysis based on the guidelines provided in Chapter 3. They also assess the need for fine-tuning the selected model.

Prompt engineers develop sample questions and extensively test the front-end application to ensure it provides relevant and accurate responses. They include prompts to prevent the system from providing incorrect or misleading information.

Outcome and Benefits

Marco and his team are thrilled with the new GenAI-based system. They can now use natural language prompts to generate insights and reports on service consumption patterns, constituent needs, and contact center operations. For example, Marco can ask questions such as the following:

> Analyze the clickstream data and social media sentiment to identify areas where we can improve the online experience for citizens seeking unemployment benefits.

> Predict the likely surge in unemployment claims based on recent economic indicators and recommend strategies to allocate resources effectively.

> Analyze call center data to identify common issues, agent performance trends, and opportunities for improving customer satisfaction.

These new capabilities dramatically improve the PSO's ability to deliver personalized and high-quality services to constituents, optimize resource allocation, and enhance overall customer experience.

8.4.3 Use Case 3: Program and Project Management

Problem

As we discussed earlier, government leaders and contractors manage large-scale programs. For example, in FY 2022, the US federal government awarded $694 billion in contracts.[8] Emma is a program manager at a PSO who spends a significant amount of time on annual, monthly, and weekly reporting and program management tasks. These tasks include drafting emails, presentation decks, and creating program management documents. She wonders, "What if there was a magic wand that could generate first drafts of most of this content!?!"

Solution

While not exactly a magic wand, GenAI is pretty close in this case and holds the potential to fulfill Emma's wish! Maria, the CTO, tells Emma that she will review and procure tools that can provide this capability. She asks her architect to evaluate the commercial off-the-shelf (COTS) products that have started to provide this capability; see Appendix E for a list of current COTS products. Her architect does a thorough requirements analysis and, using the guidelines in Chapter 3, implements a solution.

Outcome and Benefits

Emma is thrilled with the new capabilities! She is now able to get through her tasks so much more quickly! Her new GenAI product can help her with several tasks: she is able to draft monthly program management reports, generate presentation decks and speaker notes, refine existing presentation decks, and summarize action items from meetings.

---

[8] https://gaoinnovations.gov/Federal_Government_Contracting/

This means she has more time to focus on more valuable program management tasks such as meeting with vendors and contractors, creating innovative solutions, and providing value to the mission. The result? Public programs that are more efficient, on time, to make a difference for the people they serve.

8.4.4 Use Case 4: Cyber Security, Threat Intelligence, and Analytics

Problem

John is the Director of Cyber Security at a PSO. His team is responsible for safeguarding the critical infrastructure and ensuring public safety across the state. However, they face several challenges, including aging systems, lack of deeper insights into cyber threats, and the constant risk of cyber attacks that could erode public trust in the department's ability to protect citizen data. Arnav, the Chief Information Security Officer (CISO) at the PSO, recognizes the need to enhance their cyber security and compliance measures. He knows that traditional methods are no longer sufficient to keep up with the ever-evolving threat landscape. Arnav is aware of the potential of GenAI and believes it could be the solution they've been looking for.

Solution

Arnav assembles a team of experts, including John, to explore the implementation of a GenAI-based solution for enhancing cyber security and compliance at the PSO. Together, they envision the following applications:

- **Threat intelligence generation**: Data scientists and ML engineers fine-tune FMs to analyze security data, including historical incidents, user activity, security logs, and network traffic. These models can generate detailed threat intelligence reports, helping John's team gain deeper insights into potential threats. Additionally, the models can simulate various threat scenarios, enabling the security team to practice and improve their incident response capabilities.

- **Anomaly detection**: Data scientists and ML engineers fine-tune FMs to detect security threats by identifying abnormal patterns in network traffic, user activity, API calls, and other data sources. For example, if the model detects unusual traffic originating from a foreign country for unemployment insurance claims, it can flag the activity for further investigation and potential blocking.

- **Phishing and malware analysis**: Data scientists create synthetic phishing emails and malware samples using FMs. These simulations help the PSO assess the effectiveness of their security measures and employee awareness. Based on the results, the team can implement targeted training programs to educate employees on recognizing and reporting such threats.

- **Incident or event simulation and response**: FMs are fine-tuned to simulate cyber attacks, security events, and incidents. This allows John's team to assess the PSO's preparedness to detect, control, mitigate, and remediate potential attacks, ensuring that they are always ready to protect constituent data and critical infrastructure.

Outcome and Benefits

Arnav is thrilled with the implementation of the GenAI-based solution. The PSO experiences a significant improvement in their cyber security posture. John's team can now proactively identify and respond to threats more effectively, reducing the risk of successful cyber attacks on their systems. The ability to simulate various scenarios and analyze data more deeply enhances their incident response capabilities, ensuring they are better prepared to safeguard public data and maintain the trust of the constituents they serve.

8.5 Conclusion

GenAI puts business intelligence, analytics, and reporting in the hands of the PSO business staff by allowing users to interact with data using natural language. GenAI can unlock deeper insights, automate tedious reporting tasks, and provide real-time analysis to enable data-driven decision-making. GenAI can also improve customer experience and compliance with government regulations.

The use cases outlined in this chapter merely scratch the surface of what is possible. As GenAI models and the surrounding tools and platforms continue to rapidly evolve, we will likely see an explosion of new applications across the government services sector. However, realizing the full potential of GenAI for analytics and reporting is not without its challenges. Scaling these solutions to handle high concurrency while ensuring cost efficiency is important. There are some important considerations around security, privacy, and responsible development of AI systems that must be addressed.

In the next chapter, we will dive deeper into these implementation considerations for deploying GenAI solutions at scale. Key topics include security best practices, leveraging cloud scalability, cost optimization strategies, and frameworks for responsible AI development and deployment. We will examine architectural patterns, processes, and governance models to help public sector organizations navigate the technical and ethical complexities inherent to GenAI. By laying out a road map for scalable, secure, and responsible adoption, we hope to equip government agencies with the ability to truly capitalize on the transformative power of GenAI for data-driven missions. The applications for smarter reporting, deeper analytics, and cognitive automation are limitless when they are implemented thoughtfully.

CHAPTER 9

Implementation, Operations, and Maintenance

So far in this book, we have covered Generative AI (GenAI) concepts, architecture considerations, and use cases within public sector organizations (PSOs). In this chapter, we will delve into implementation considerations that enable responsible, secure, and successful deployment of GenAI solutions. Our goal is to equip readers with specific technical considerations and nuances for GenAI implementation. In this chapter, we will dive deep into the following topics:

- **Anatomy of a GenAI application**: We review this from a technical perspective. This includes tools and technologies across six components: security, application, orchestration, model, data, and infrastructure.

- **Foundation Model Operations (FMOps)**: This is a methodology that focuses on the processes, techniques, and best practices applied for managing the model and data components of a GenAI application.

© Sanjeev Pulapaka, Srinath Godavarthi and Dr. Sherry Ding 2024
S. Pulapaka et al., *Empowering the Public Sector with Generative AI*,
https://doi.org/10.1007/979-8-8688-0473-1_9

- **Scalability and performance management**: These are aspects that can address increasing demands and workloads while delivering consistent and reliable results. We discuss this for all components except security.

- **Security and privacy considerations**: These are some of the most important aspects for any application in the public sector, especially for GenAI-based applications. We therefore cover security and privacy aspects separately in detail.

- **Cost optimization**: Costs are an important consideration for PSOs operating within constrained budgets. We review various costs incurred throughout the GenAI life cycle and examine approaches to optimizing the costs and maximizing the value of GenAI investments.

9.1 Anatomy of a GenAI Application

To understand the performance considerations, let's start by reviewing a typical GenAI application from a technical point of view. Figure 9-1 illustrates the anatomy of a GenAI application with multiple components. Each component represents a set of technologies fulfilling certain key functionalities.

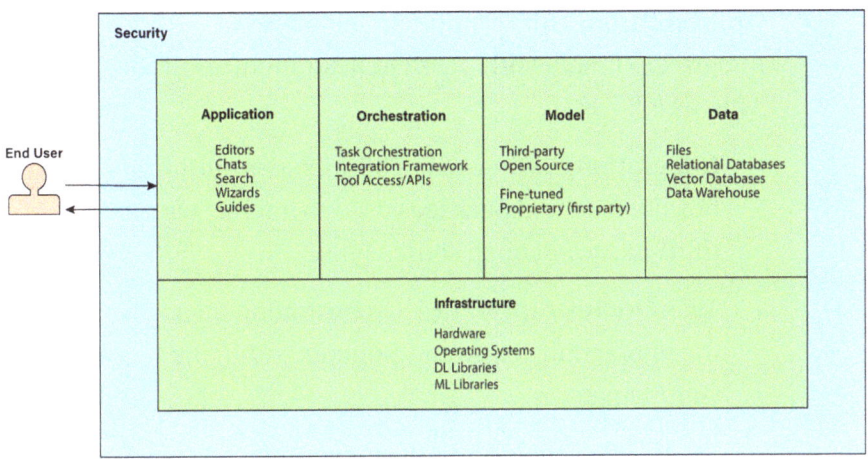

Figure 9-1. *Anatomy of a GenAI Application*

Let's review each of the components starting with the Security layer and then working our way from Application to Infrastructure layers:

- **Security**: This component represents the security measures and controls that are implemented across the entire GenAI application stack to ensure data privacy, security, and compliance.

- **Application**: This component represents the front-end applications that end users use to leverage GenAI capabilities. This includes web and mobile applications and user interfaces. This component includes technology and software for various use cases we discussed earlier such as contract generation, synthetic image generation, chatbots and search, image generation, and comment summarization.

- **Orchestration**: This component includes the Integration Framework, Task Orchestration, and Tools/APIs, as follows.

    - **Integration Framework**: This includes out-of-the-box libraries and packages used to interact with third-party and open source FMs.

    - **Task Orchestration**: Task Orchestration manages the interaction across components within the GenAI anatomy.

    - **Tools/APIs**: Tools and APIs for various tasks, such as document management, calculations, SQL, image formatting, and so on.

- **Model**: This component consists of the various FMs that power GenAI applications. See Chapter 3 for more details about each of these. The model could be one of several types of models available:

    - **Third party**: These are pre-trained Large Language Models such as Claude, Gemini, and GPT, developed by organizations such as Anthropic, Amazon, Google, OpenAI, and so on.

    - **Open source FMs**: These include Llama, Hugging Face, Mistral, and so on, available on platforms such as Hugging Face, etc.

    - **Fine-tuned (domain specific)**: These are FMs that have been further fine-tuned for specific use cases or domains.

    - **Proprietary (first party)**: These FMs are proprietary models developed and trained by the organization itself. Developing an FM is a complex activity which needs a lot of expertise and specialized resources. We expect that most PSOs will not need to develop these models. Therefore, we don't discuss considerations for these models in this chapter.

- **Data**: These are all the technologies associated with data and metadata storage such as vector databases, files, relational databases, data warehouses, and so on.

- **Infrastructure**: This component involves the hardware for running all the other components. The infrastructure includes operating systems such as Linux or Windows; deep learning libraries such as TensorFlow and PyTorch that are necessary to run FMs; end-to-end ML platforms such as Amazon SageMaker, Hugging Face Deep Learning Containers on AWS, and Azure Machine Learning Service; and Google Vertex AI. For PSOs that opt for open source models, this could also include open source model inference frameworks such as BentoML and Ray and any other libraries required for running the components.

This approach allows us to clearly view the technical aspects of GenAI applications by separating concerns across different application layers.

9.2 Foundation Model Operations (FMOps)

FMOps stands for Foundation Model Operations. It draws inspiration from MLOps, Machine Learning Operations, which itself is a derivation from a methodology known as DevOps, Development and Operations! Broadly speaking, these terms are methodologies that define tools and best practices to efficiently manage technology implementations and operations. Their goal is to improve the overall efficiency of, and reduce the risk associated with, the use of technology in organizations.

The concept of DevOps was introduced to efficiently manage software implementations. Therefore, in the case of a GenAI application, we can infer that DevOps could be used to manage most of the components in our GenAI application with the exception of the Models component. The reason is that activities associated with the Models component are quite unique and need special attention. MLOps was introduced to manage the specific activities associated with training, testing, deploying, and managing machine learning models. However, the activities associated with FMs are not quite the same as traditional ML. For example, many organizations will not need model training activities if they procure pre-trained FMs. Therefore, you now have the field of FMOps which focuses on the specific activities associated with FMs.

Figure 9-2 illustrates the types of activities associated with FMs. Note that we are not including pre-training activities of FMs. As mentioned earlier, pre-training is a complex activity, and a discussion of that subject is out of scope for this book.

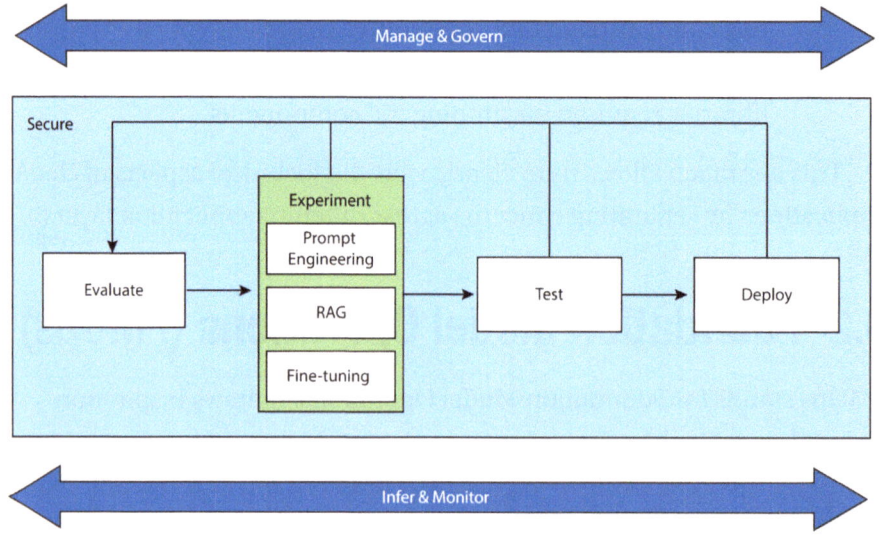

Figure 9-2. *Foundation Model Operations*

So, one might ask, what's the difference between the FMOps and the GenAI life cycle introduced in Figure 3-2 in Chapter 3? The main difference lies in scope and focus. The GenAI life cycle is the broader framework encompassing the entire development and management process of developing a GenAI application. It includes project and product management, management of users, strategies, business cases, and so on. FMOps, on the other hand, focuses on tools and best practices to manage and streamline FM-specific activities in a production environment. Let's go through each of the activities mentioned in Figure 9-2 to understand them in more detail. Note that some tools and best practices apply to multiple activities.

9.2.1 Manage and Govern

This includes tools and best practices to manage the entire life cycle of FMs, from model pre-training, fine-tuning, and versioning to deployment. FMOps provides tools to set up a model repository with automated processes and APIs for updating, saving, and versioning models. The model repository provides a model catalog with detailed metadata about each model to allow downstream users to pick the model of choice. It also allows users to switch between different models and model providers via a unified interface.

9.2.2 Evaluate

In Chapter 4, we discussed evaluation techniques for measuring the effectiveness of FMs as well as techniques such as Retrieval-Augmented Generation (RAG). FMOps recommends tools and best practices to automate the evaluation process. This includes workflows for evaluation when selecting a model for a use case. It also includes evaluation when a model is moved into testing and production.

9.2.3 Experiment

Effective utilization of FMs relies heavily on crafting appropriate prompts and guiding the model's generation process. FMOps advocates prompt libraries, catalogs, and use of automated tools for prompt management, version control, testing, and optimization to improve model outputs and align them with desired behaviors. These tools allow users to track and visualize prompting and fine-tuning experiments in real time. They also work hand in hand with model management tools to enable development, testing, and rollout of optimal prompts, parameters, models, and model providers for a given use case. FMOps also recommends using automation and tools for RAG and fine-tuning. For RAG, the tools would assist with evaluation using some of the metrics described in Chapter 4. Fine-tuning tools include the capability to train version models and use the same evaluation techniques mentioned for measuring effectiveness earlier to determine the best models.

9.2.4 Test

We discussed model evaluations as well as measuring effectiveness of RAG and fine-tuning approaches in the Experiment phase. However, in a production environment, having a separate end-to-end activity for testing the application that leverages the FM is essential. To some extent, this may also be covered under the DevOps methodology. Testing should ideally be carried out using a series of standardized test cases and scenarios and should cover FM performance, latency, accuracy, robustness, toxicity, security, integration with other tools, and compliance with regulations. The tools used here could be standardized testing platforms such as Jenkins or cloud-based tools such as AWS CodePipeline and Azure DevOps.

9.2.5 Deploy

PSOs have three broad choices for GenAI application deployment:

1. Use comprehensive cloud-based services such as Amazon Bedrock, Azure OpenAI Service, and Google's Vertex AI that provide access to a number of open source and proprietary FMs. In this case, the cloud services provide APIs for model deployment.

2. Use model hubs such as Amazon SageMaker Jumpstart, Azure Model Catalog, and Google Model Garden to deploy open source or proprietary FMs to cloud-based infrastructure. In this case, the PSO has to manage deployments, although the infrastructure itself is provided by the cloud service provider. However, the cloud provider may offer libraries, packages, and necessary guidance to reduce complexity of deployments.

3. Directly deploy open source models. In this case, PSOs can deploy open source models such as Llama, Jurassic-1 Jumbo, and CLIP, many of which are available on Hugging Face, either on infrastructure in the cloud or on-premises. These deployments in particular can be highly complex and challenging due to the size, computational requirements, and potential impact. FMs often require specialized hardware infrastructure, such as GPU clusters or TPU pods, due to their massive computational requirements. To reduce the computational footprint and deployment complexity, techniques such as knowledge distillation or model compression can be employed.

These methods create smaller, more efficient models that approximate the behavior of the larger FM, making deployment easier. In scenarios where data privacy or regulatory constraints are a concern, federated learning[1] or decentralized deployment strategies can be used for activities such as fine-tuning. In these approaches, the FM is deployed across multiple distributed locations without centralizing the data. Appendix F provides FMOps tools that PSOs can use to manage deployments when using this option.

FMOps activities for deployment also include consideration and tools for different types of rollout options. Let's review these:

- **Approvals**: Before rollout, a workflow needs to be established for model approvals. This is especially necessary for PSOs for compliance and regulatory reasons.

- **Incremental rollout**: Instead of deploying the FM (pre-trained or fine-tuned) all at once, this strategy involves rolling it out gradually to a subset of users or use cases. This is important because providers keep releasing updates and versions of models. This allows for monitoring, feedback collection, and adjustments before a broader release. Techniques such as canary deployments, where a small percentage of traffic is routed to the new model, can be employed.

- **Staged deployment**: This is similar to incremental rollout, but with distinct stages or phases. For example, the FM might first be deployed internally for testing, then to a limited set of beta users, and finally to the broader public or production environment.

---

[1] https://research.google/blog/federated-learning-collaborative-machine-learning-without-centralized-training-data/

- **Blue-green deployment**: In blue-green deployment, you create two separate but identical environments. For deployments, you deploy to the first environment, complete testing, and then deploy the model to the second environment.

The specific deployment and rollout strategy chosen will depend on factors such as the intended use case, computational resources available, data privacy and regulatory requirements, and the organization's risk tolerance and governance processes.

9.2.6 Infer and Monitor

This phase consists of inference using deployed FMs as well as monitoring of the inferences. As explained in Chapter 1, inference is using the FMs to generate outputs. In other words, this is where end-user applications, such as chatbots, invoke the deployed FM for various tasks, including search, content generation, and summarization.

Monitoring the outputs from the FM is an important activity given the nature of FMs. FMOps recommends tools and processes to continuously monitor outputs of FMs to measure effectiveness using the metrics described in Chapter 4. The automation could also include raising alerts when discrepancies are found or thresholds exceeded.

FMOps also recommends tools to continuously examine responsible and ethical use of models, aligning with organizational policies and regulatory requirements.[2] For example, use guardrails to ensure automatic flagging and correction of prompts and outputs to remove undesirable and

---

[2] www.whitehouse.gov/briefing-room/presidential-actions/2023/10/30/executive-order-on-the-safe-secure-and-trustworthy-development-and-use-of-artificial-intelligence/

toxic content. In addition to these tools, you also need tools for real-time logging and error tracking of FM invocations to evaluate latency, time-outs, and metrics such as queries per second and tokens per second.

9.2.7 Secure

Given the potential sensitivity of data used by FMs and the risk of privacy violations or misuse, FMOps emphasizes implementing robust security measures, such as access controls, data encryption, and secure communication protocols to protect sensitive information and maintain user privacy. We will go into more detail on security later in Section 9.5.

9.3 FMOps Benefits

Why is FMOps important? And what do PSOs specifically gain from it? Imagine there are a number of teams across a PSO that are working with GenAI applications. As a PSO, you need to ensure a number of things:

- **Standardization and consistency**: There are a plethora of choices and different types of FMs available for different purposes. Therefore, there needs to be some level of consistency across all teams involved with building GenAI applications. This ensures consistent management across different GenAI applications and projects, leading to fairer and more reliable outcomes for constituents.

- **Governance**: You need to control access to FMs, preventing unauthorized modifications or misuse. You also need version control to track different versions of models for easy rollbacks if needed, ensuring model stability and mitigating risks. Comprehensive documentation clearly outlining the model's intended function and applications is also a must to enable audits and regulatory reporting.

- **Responsible AI, monitoring, and evaluation**: As mentioned throughout this book, outputs from FM need to be reviewed in a consistent manner. This means you need to perform checks for errors, latency, bias, fairness, and security vulnerabilities throughout the GenAI life cycle (illustrated in Figure 3-1). The earlier in the life cycle you catch issues, the easier and faster it is to fix them. For example, an issue with a deployed chatbot used by hundreds of people across the organization not only is more expensive to resolve but also results in loss of trust in the application.

The tools and capabilities for FMOps that PSOs need to consider may vary based on the type of deployment. Currently, cloud providers such as AWS, Azure, and Google perhaps provide the most comprehensive platforms for end-to-end management. However, PSOs that opt for self-deployment using open source models need to consider options provided in Appendix F. Given the rapidly developing space, newer tools and functionalities are bound to emerge. As the adoption of FMs and GenAI continue to grow, FMOps will play a crucial role in enabling the reliable, scalable, and trustworthy operationalization of these powerful technologies.

9.4 Performance, Reliability, and Scalability of GenAI Applications

Building an application for a proof of concept is vastly different from building a production-ready application. For example, a user interacting with a GenAI chatbot may not want to wait 30 seconds to a minute for a response to simple questions. The difference comes down to factors such as performance, scalability, and reliability.

In this section, we discuss important considerations for these factors for all the components in the GenAI anatomy, shown in Figure 9-1, except security, which we'll discuss separately in Section 9.5. We discuss considerations for PSOs looking to self-host open source FMs on-premises or in the cloud as well as PSOs looking to use third-party models in the cloud. Note that our intent is to focus more on the specific nuances relevant to GenAI for each component.

9.4.1 Application

From an application standpoint, enhancing performance, scalability, and reliability is no different from any other web or mobile application. For the front end itself, consider techniques including code minification, image optimization, asynchronous processing, and Content Delivery Networks (CDNs). The front end should be complemented by a scalable microservices and event-driven architecture. Techniques such as load balancing, use of in-memory data stores, and caching further help optimize performance. Additionally, redundancy in critical components, robust error handling, and monitoring with alerts all enhance reliability. Again, these approaches are the same as any other web or mobile applications in production.

9.4.2 Orchestration

While open source orchestration frameworks, such as LangChain and LlamaIndex, are valuable for testing and proof of concepts, PSOs aiming for production environments should explore the tools listed in Appendix F. These frameworks offer libraries specifically designed to enhance the performance of orchestrators.

One key challenge lies in scaling the embedding generation process for large datasets, which requires significant memory and processing power. Frameworks, such as Ray, can address this by parallelizing the process, significantly reducing processing time. Additionally, utilizing cloud-based platforms and services can further aid in scaling your GenAI application.

Major cloud providers are increasingly offering orchestration capabilities. For example, AWS provides knowledge bases for Amazon Bedrock, and Google offers Vertex AI Search and Conversation.

So how can we measure the performance of the Orchestrator layer? Depending on the type of use case, a number of metrics can be used. For example, if the Orchestrator is executing a RAG pipeline, the following metrics are used:

- **Indexing metrics**: These measure recall, indicating the number of relevant results returned by the vector search algorithm.

- **Retrieval metrics**: This includes Normalized Discounted Cumulative Gain (NDCG), a metric to assess ranking quality based on how much relevant content is captured (recall) and how many retrieved results are truly relevant (precision).

- **Generation metrics**: These evaluate the faithfulness, answer relevance, and overall quality of generated responses using metrics similar to Sensibleness and Specificity Average (SSA).

Tools such as Ragas and Tonic offer readily available libraries to simplify the evaluation process. Ragas, for example, is a powerful library that analyzes performance by collecting the input, output, and context and deriving metrics for both the retriever and generator components. It measures context precision and recall for retrieval and faithfulness and answer relevancy for generation.

9.4.3 Model

The deployment approach described earlier in the section under FMOps has a large say in the performance implications of FMs. Cloud-based services such as Amazon Bedrock, Azure OpenAI Service, and Google's Vertex AI handle all the engineering needed to optimize the performance of models. However, PSOs evaluating self-hosting of open source models can consider the following strategies:

- **Knowledge distillation**: Train a smaller model to mimic the predictions of the larger Foundation Model. The smaller model becomes faster for inference while retaining good accuracy.

- **Batching**: When possible, group multiple inference requests into batches for the model to process simultaneously. This can improve efficiency on some hardware architectures.

- **Asynchronous inference**: Offload model inference to background threads to avoid blocking the main application thread. This keeps the UI responsive while waiting for model results.

- **Model caching**: Cache frequently used model outputs to avoid redundant inference calls. This can significantly improve performance for repetitive tasks.

9.4.4 Data

Performance at the data level is crucial for any application. Regardless of how well optimized the rest of your architecture is, storage, processing, and retrieval of data from databases can severely impact the performance, scalability, and reliability of your application. Data storage and retrieval are especially critical for techniques such as RAG and fine-tuning. A thorough examination of advanced retrieval strategies is outlined in the paper "Retrieval-Augmented Generation for Large Language Models: A Survey."[3] Let's review some of the most important considerations and techniques for optimizing performance of data storage, processing, and retrieval, including some techniques outlined in this paper:

- **Distributed file and processing systems**: Explore distributed file systems such as HDFS (Hadoop Distributed File System), large-scale data processing systems for large datasets such as Apache Spark, and distributed search and analytics platforms such as OpenSearch. These systems distribute data across multiple machines, enabling parallel access and improved performance for GenAI applications that process massive datasets.

- **Data life cycle management (DLM)**: Implement DLM strategies to optimize data storage utilization. DLM involves automatically archiving or deleting less frequently accessed data, freeing up space on high-performance storage for active training data. This keeps storage costs under control while maintaining performance for critical tasks.

---

[3] http://arxiv.org/pdf/2312.10997

- **Data compression**: Explore data compression techniques suitable for your GenAI application's data format. Compression reduces storage requirements and can improve access times, especially for frequently accessed datasets.

- **Data cleaning and preprocessing**: Ensure data is clean and free of errors including inconsistencies and irrelevant information. Techniques such as text normalization, entity recognition, and duplicate removal can be used.

- **Data filtering and selection**: Focus on high-quality data relevant to the specific task or domain. This improves the effectiveness of RAG retrieval and fine-tuning by providing the model with the most pertinent information.

- **Indexing optimization**: In RAG, the indexing phase involves processing documents, segmenting them, and converting them into embeddings for storage in a vector database. In the indexing phase, chunking strategies that determine the granularity of retrieved data (e.g., tokens, sentences, documents) is very important. This involves finding the right balance between context length and the ability to capture meaning. Metadata attachments such as timestamps or author information can be used to filter retrieval results. Building a hierarchical structure for documents can improve retrieval efficiency.

- **Embedding models**: As discussed in the RAG section in Chapter 4, embeddings are used to calculate similarity between the query (prompt) and document chunks for retrieval. Popular choices include sparse encoders (BM25) and dense retrievers (Bidirectional Encoder Representations from Transformers (BERT)-based models). Recent advancements include embedding models such as AngIE, Voyage, and BGE, which benefit from multitask training. Cloud providers may provide proprietary embedding models such as Amazon Titan and Cohere Embed.

The best embedding model depends on the specific use case. Hybrid approaches combining sparse and dense retrieval can leverage the strengths of both methods. Fine-tuning embedding models on domain-specific datasets can improve performance, especially for specialized fields with unique jargon.

9.4.5 Infrastructure

Just as a high-performance car needs a strong engine and smooth roads, GenAI applications rely heavily on the infrastructure component of the GenAI application anatomy. The infrastructure provides immense processing power to (1) train and run complex models, (2) store and manage massive datasets, (3) scale to adapt to growth, (4) optimize performance for faster results, and (5) ensure reliability and security for sensitive data – all crucial factors for GenAI to function effectively and reach its full potential.

PSOs using comprehensive cloud-based services such as Amazon Bedrock, Azure OpenAI Service, and Google's Vertex AI do not need to concern themselves with this aspect as the service takes care of all the heavy lifting required behind the scenes. However, PSOs that don't use these services need to review the following considerations for a high-performing, reliable, and scalable GenAI application:

- **Graphics processing units (GPUs) or similar accelerators**: GenAI models are computationally intensive, and using machines that offer GPUs significantly speeds up inference. However, with the advent of GenAI, GPUs are becoming increasingly difficult to obtain. PSOs can also explore other options such as AWS Inferentia, a high-performing chip designed by AWS for high-performance inference predictions.

- **Memory (RAM)**: Ensure sufficient RAM to handle the model and data being processed. Insufficient RAM can lead to bottlenecks.

- **Storage**: Use a combination of solid-state drives (SSDs) for fast data access during training and inference and hard disk drives (HDDs) for larger datasets that aren't constantly accessed. Plan for future growth: GenAI models and datasets can be large, so ensure there is enough storage to accommodate them.

- **Distributed storage**: Explore distributed storage systems across multiple machines for larger datasets and redundancy.

- **Deep learning frameworks**: Leverage frameworks such as TensorFlow and PyTorch that support distributed training across multiple GPUs or machines for faster training of complex models. Explore framework-specific optimizations such as model pruning or quantization to reduce computational requirements.

- **Containers**: Explore containerization technologies such as Docker to package model dependencies and to ensure consistent performance across environments.

- **Open source frameworks**: For open source models, consider tools listed in Appendix F for optimized model serving and efficient inference.

- **Cloud-based ML platforms**: PSO leveraging cloud platforms can explore end-to-end ML platforms such as SageMaker, Azure Machine Learning, or Vertex AI. These platforms provide most of the capabilities listed earlier, such as automatic model optimization, resource management, and scaling. They also provide capabilities to automatically adjust resources based on workload, optimizing cost and performance. This includes features such as auto-scaling or managed versions of container orchestration tools such as Kubernetes to automatically scale compute resources up or down based on demand.

As we conclude this section, it is important to note that achieving optimal performance, scalability, and reliability for GenAI applications is a multifaceted endeavor that requires careful consideration of various components within the GenAI application anatomy. From optimizing front-end and back-end architectures to leveraging advanced data management techniques, orchestration frameworks, and efficient model deployment strategies, a holistic approach is necessary to ensure a seamless user experience. Because of the complexity involved, harnessing the power of cloud-based services can particularly help PSOs accelerate their journey into GenAI.

9.5 Security and Privacy

In previous chapters, we discussed how GenAI can be used for a wide range of applications in the public sector. However, without adequate security- and privacy-related controls, PSOs face an increased level of risk of privacy-related attacks. PSOs can leverage resources such as MITRE Adversarial Threat Landscape for Artificial-Intelligence Systems (ATLAS) knowledge base,[4] Open Web Application Security Project(OWASP) AI Security and Privacy Guide,[5] and NIST's Artificial Intelligence Risk Management Framework (AI RMF 1.0)[6] for details on how to handle the various security- and privacy-related risks associated with AI in general. Note that these may apply to AI in general and are not necessarily specific to GenAI. Given this is a relatively new field, a lot of organizations are still in the process of establishing standards for mitigation of GenAI-specific risks.

In this section, we distill the most critical risks and mitigation strategies for each component of the GenAI application anatomy represented in Figure 9-1. As in previous sections, considerations vary between self-hosting and cloud-based hosting strategies. Cloud-based providers offer a lot of features to ensure security and privacy out of the box. However, they do adopt a shared responsibility model for security, and there may be some aspects of security for which the PSO needs to be responsible.

9.5.1 Application

- **Content filtering**: Use content filters to evaluate prompts. Ideally, ensure that any Personally Identifiable Information (PII) is not presented to the

---

[4] https://atlas.mitre.org/

[5] https://owasp.org/www-project-ai-security-and-privacy-guide/

[6] https://nvlpubs.nist.gov/nistpubs/ai/nist.ai.100-1.pdf

FM. Censure prompts that violate industry, regulatory, and organizational standards.

- **Implement data uploading controls**:[7] Control the data that is uploaded to your application. For example, implement data loss prevention (DLP) solutions to detect and prevent the upload of data that violates organization policies.

- **Prompt injection and jailbreak**: Implement mechanisms to analyze prompts for attempts to discover or override system instructions, which could lead to the model behaving maliciously. This could involve techniques such as prompt filtering, watermarking, or other defensive measures.

- **Sensitive data phishing**: Analyze prompts for attempts to gain access to sensitive information, such as trade secrets, financial data, or personal information. Implement appropriate access controls and monitoring mechanisms to detect and prevent such attempts.

9.5.2 Orchestration

- **API keys**: Orchestrators access models using API endpoints. Most APIs use API keys to authenticate users and authorize access to specific functionalities. Proper storage and management of API keys are essential. Avoid hardcoding API keys in code or scripts, as this can lead to unauthorized access if the code is compromised. Instead, store API keys securely

---

[7] https://aws.amazon.com/blogs/security/securing-generative-ai-applying-relevant-security-controls/

in a centralized key management system or use
environment variables.

- **Access control**: APIs should have robust access control
  mechanisms in place to restrict access based on user
  permissions and roles. Role-based access control
  (RBAC) and attribute-based access control (ABAC) are
  common approaches to implement granular access
  control policies.

- **Provider security**: When integrating with third-party
  FM providers, it is crucial to assess their security
  practices and reputation. Look for adherence to
  relevant compliance standards (such as ISO 27001
  or SOC 2) and secure coding practices. Verify
  their security measures and stay updated on any
  vulnerabilities reported in their APIs.

- **Data sharing**: Be mindful of the data you share with
  third-party FM providers and understand their data
  privacy practices. Avoid sharing sensitive information
  or PII unless absolutely necessary and ensure that
  proper data protection measures are in place.

- **Integration**: Orchestrators use agents to connect to
  a number of tools within your organization. Ensure
  that the identity of the application user is propagated
  correctly to the systems so that the source systems are
  not compromised in any manner.

9.5.3 Model

- **Model access**: Limit direct access to the FM API endpoints, whether you self-host the FM or consume the FM from a cloud provider. Access should be controlled based on users' needs and profiles. For self-hosting, consider setting up access through centralized platforms as opposed to allowing users to directly query from endpoints. Control access to the fine-tuned model artifacts and endpoints according to the classification of the data used in the fine-tuning.

- **Content filtering**: Analyze outputs of the FMs for inappropriate content. Develop a set of rules and benchmarks to continuously test and evaluate the quality of the output.

- **Denial of service (DOS)**: Prevent behavior that may lead to stalling the model for legitimate uses. This could involve rate limiting, resource monitoring, and other techniques to ensure fair and equitable access to the model.

- **Anomalous behavior**: Scan for general anomalous access or prompt content that warrants additional inspection. This could involve machine learning–based anomaly detection, user behavior analytics, or other techniques to identify and investigate suspicious activity.

- **Model source verification**: Verify that the model artifacts are from a trusted source with digital signature verification and implement scanning the model

251

artifacts for vulnerabilities. If you're consuming the FM from a cloud provider, verify that the cloud provider implements these controls.

- **Red teaming:**[8] Conduct regular red team exercises to identify potential vulnerabilities and weaknesses in your GenAI system. This can help uncover blind spots and inform continuous improvement of your security posture.

9.5.4 Data

- **Encryption**: Implement encryption for data used for techniques such as RAG and fine-tuning. Also, ensure encryption is in place in transit to protect the confidentiality and integrity of sensitive data. Use strong encryption algorithms (such as AES-256) and secure key management practices. Encrypt model artifacts, whether pre-trained or fine-tuned.

- **Access controls**: Apply granular access controls to restrict access to sensitive data based on user roles and permissions. Regularly review and revoke unnecessary data access privileges.

- **Data masking**: Use data masking techniques to obfuscate sensitive information in non-production environments, such as development or testing environments, where full data access may not be necessary.

---

[8] https://hbr.org/2024/01/how-to-red-team-a-gen-ai-model

9.5.5 Infrastructure

- **Network security**: Implement firewalls and intrusion detection/prevention systems (IDS/IPS) to monitor network traffic for suspicious activity. Configure these tools to alert on and block unauthorized or malicious traffic patterns. Consider using private connections to cloud providers instead of routing traffic through the Internet.

- **Network segmentation**: Segment your network to isolate workflow orchestration systems and GenAI components from other critical resources, minimizing the potential impact of a breach. Use virtual private networks (VPNs) or software-defined perimeters to secure communication between different network segments.

- **System hardening**: Keep operating systems and software on your infrastructure patched and updated to address known vulnerabilities. Implement a robust patch management process to ensure timely and consistent application of security updates.

- **Operating system authentication and authorization**: Enforce strong authentication and authorization mechanisms for accessing operating systems and infrastructure components. Use multifactor authentication (MFA), strong password policies, and principle of least privilege for user accounts and service accounts.

- **Security monitoring and logging**: Implement comprehensive security monitoring and logging capabilities to detect and investigate potential security incidents. Leverage security information and event management (SIEM) solutions to collect and analyze logs from various sources, including network devices, servers, applications, and security tools.

- **Incident response and disaster recovery**: Develop and regularly test incident response and disaster recovery plans to ensure resilience and business continuity in the event of a security breach or other disruptive events.

- **Regular security assessments**: Conduct regular security assessments, such as penetration testing, vulnerability assessments, and code reviews, to identify and remediate potential vulnerabilities proactively.

By implementing these security and privacy measures across the different components of your GenAI application, PSOs can mitigate risks and build trust in their solutions, enabling the safe and responsible use of GenAI.

9.6 Cost Optimization

GenAI applications offer tremendous opportunities for innovation and automation. However, these applications can incur significant costs over time if they are not well architected, built, managed, and optimized.

In this section, we analyze infrastructure, hardware, and tooling costs, as well as personnel and human expertise costs related to implementing a GenAI application. We divide the costs into two types: upfront and ongoing. Upfront costs primarily occur during the development stages of a GenAI application. Ongoing costs are generated continuously while the application is in operation. We wrap up the discussion by reviewing various strategies to optimize expenses, using relevant examples from the public sector.

9.6.1 Infrastructure, Hardware, and Tooling Costs

These are costs associated with the platform and tools used throughout the GenAI life cycle illustrated in Chapter 3, Figure 3-2. Table 9-1 illustrates the costs for each phase of the life cycle and defines whether each cost is upfront or ongoing.

Table 9-1. *GenAI Life Cycle Costs*

Life Cycle Stage	Cost Categories	Upfront or Ongoing
Business Problem Definition and Planning	Data Acquisition (external data sources, internal data retrieval) for analysis	Upfront
Data Collection and Processing	Data Acquisition (purchase/license data)	Upfront
	Data Storage (cloud storage)	Ongoing
	Data Processing (cleaning, preparation, transformation)	Upfront
FM Evaluation and Selection	Cloud Services (model evaluation and testing)	Upfront
FM Fine-Tuning	Hardware (computing resources – GPUs, servers)	Upfront
	Software Tools (model training frameworks)	Upfront

(*continued*)

Table 9-1. (*continued*)

Life Cycle Stage	Cost Categories	Upfront or Ongoing
	Cloud Services (model fine-tuning)	Upfront
Application and Orchestration Layer Development	Software Development (APIs, connectors, user interfaces)	Upfront
	Cloud Services (Application and Orchestration Layer Development)	Upfront
	Third-Party Tools (orchestration and integration)	Upfront
Testing, Validation, Monitoring, and Auditing	Cloud Services (testing and performance monitoring)	Ongoing
	Auditing Tools (bias detection, model output analysis)	Upfront
Production Deployment	Infrastructure (servers, network resources)	Upfront
	Cloud Services	Ongoing
	Scalability (infrastructure scaling for user load and data volume)	Ongoing
	Disaster Recovery (infrastructure redundancy)	Upfront
Hardware Licensing Fees	Licensing fees for different kinds of hardware if self-hosted	Ongoing
Software Licensing Fees	Licensing fees for different kinds of software if self-hosted	Ongoing

(*continued*)

Table 9-1. (*continued*)

Life Cycle Stage	Cost Categories	Upfront or Ongoing
Continuous Monitoring, Auditing, and Fine-Tuning	Cloud Services (monitoring and data processing)	Ongoing
	Application Updates	Ongoing
	Model Updates (retraining, updates from FM provider)	Ongoing
	Costs associated with bias mitigation, fairness audits	Ongoing
	Costs associated with data privacy and security compliance	Ongoing

Let's review these costs using the development of a chatbot as an example.

Upfront Costs

Upfront costs primarily occur during the development stages of a GenAI application. However, this may also be a function of the type of deployment. For example, for PSOs that use self-hosting models, there may be upfront costs to acquire the infrastructure and software.

After defining the chatbot's scope and functionality, the PSO may leverage a RAG solution, which requires collecting and preparing data to build a vector database.

Acquiring and curating domain-specific data, such as constituent inquiries or policy documents, into the GenAI application repository can be expensive, depending on the complexity and level of processing required. In addition, ingesting the processed data into the vector database incurs a data embedding and ingestion cost.

Various pre-trained language models need to be tested to find the best fit for performance expectations, which involves configuring parameters, designing prompts, and iterative experimentation – all aspects that generate costs in compute resources such as GPUs.

If none of the pre-trained models meet the needs, fine-tuning may be necessary. This process involves additional data collection, processing, and iterative experiments to find the optimal fine-tuning strategies and hyperparameters, further increasing upfront costs.

After selecting the language model, the orchestrator and front-end interface need to be developed, incurring tooling costs. Finally, deploying the end-to-end application in production requires infrastructure costs for testing, ensuring workload handling, error management, and concurrency.

Ongoing Costs

Ongoing costs are generated continuously while the application is in operation. Again, the timing of the costs may depend on the deployment model. For example, infrastructure costs for cloud-based providers could be classified as ongoing.

The primary ongoing cost from an FM perspective is inference expenses. These are the costs incurred each time a user query (prompt) hits the FM. If the deployment is cloud based, this could be based on the number of tokens processed by the FM or based on a fixed throughput capacity. As the user base grows and demand increases, these costs will escalate.

In a RAG solution, the vector database also generates ongoing costs for vector storage, search, and query operations. Maintaining the application requires continuous monitoring and periodic performance analysis, incurring costs for tooling and hardware.

If the application's performance degrades, the agency must investigate the root cause and determine whether prompt engineering techniques or a better model is required, generating additional experimentation costs.

Furthermore, updating the vector database to keep information current will incur recurring costs associated with data collection, preparation, embedding, and ingestion.

Last but not least, costs are incurred with implementation of security and privacy tools for bias mitigation, fairness audits, and security compliance.

9.6.2 Human Expertise Costs

Building a GenAI application requires a team of skilled professionals with expertise in data engineering, machine learning, software development, project management, and more as described in Chapter 3. Attracting and retaining such talent can be expensive and challenging for PSOs. Ongoing training and professional development may also be necessary to keep the team up to date with the latest GenAI advancements. Table 9-2 shows the costs incurred across the different phases of the GenAI life cycle.

Table 9-2. *GenAI Staffing Costs*

Life Cycle Stage	Human Resource Costs	Upfront or Ongoing
Business Problem Definition and Planning	Stakeholder engagement and requirements gathering (project managers, domain experts)	Ongoing
Data Collection and Processing	Data acquisition (data engineers)	Upfront
FM Evaluation and Selection	Research and evaluation of Foundation Models (data scientists)	Upfront
FM Training and Fine-Tuning	Data scientists and ML engineers for model fine-tuning	Upfront
Application and Orchestration Layer Development	Software developers for building APIs, connectors, and user interfaces	Upfront
Testing, Validation, Monitoring, and Auditing	ML engineers for testing, validation, and ongoing monitoring	Ongoing
Production Deployment	System administrators and IT personnel for infrastructure management	Ongoing
Continuous Monitoring, Auditing, and Fine-Tuning	Data scientists and ML engineers for ongoing monitoring, performance analysis, and potential model adjustments	Ongoing
Responsible AI Practices	Personnel with expertise in bias mitigation, fairness audits, and ethical considerations	Ongoing
Legal and Regulatory Compliance	Legal counsel and compliance specialists to ensure data privacy and security adherence	Ongoing

Let's take a look at these costs using the same example of developing a chatbot.

Upfront Costs

These are costs of the personnel involved in activities such as gathering requirements, preprocessing relevant data from internal or external sources, researching and evaluating different FMs, fine-tuning the model if required, building APIs and user interfaces, and prompt engineering.

Ongoing Costs

These are costs associated with continuously testing and validating the chatbot's performance and outputs, ongoing monitoring to identify potential biases or issues requiring adjustments, managing the infrastructure, potentially fine-tuning or adjusting the model based on user interactions and feedback, requiring data scientists and machine learning engineers, and ongoing efforts by legal counsel and compliance specialists to ensure the chatbot adheres to data privacy regulations and security standards within the public sector.

9.6.3 Considerations to Optimize the Cost

As we've seen earlier, there are a number of different costs that are incurred with a GenAI implementation. Optimizing these costs is crucial to the success of these projects. Here, we discuss some considerations to optimize the cost. By making the right choices at the onset of your GenAI effort, you can better control the upfront and ongoing costs for your GenAI application.

Project Planning

- **Focus on high-impact use cases**: Prioritize projects with clear ROI potential, such as automating repetitive tasks, personalizing public services, or optimizing resource allocation. This ensures GenAI delivers significant benefits and cost savings.

- **Phased implementation**: Break down the project into manageable phases. This allows for better cost control, resource allocation, and adaptation based on learnings from each phase.

Cost-Effective Solutions

- **Leverage open source models**: Use pre-trained, open source models whenever possible to significantly reduce upfront development and training costs. However, consider the trade-off with functionality because proprietary models may offer better options.

- **Data sharing and collaboration**: Explore partnerships with other PSOs to share data resources and potentially collaborate on building and maintaining GenAI models.

- **Hybrid team structure**: Combine in-house talent with external expertise. Build a core team with essential skills and outsource specific tasks such as data labeling or model training to specialized vendors when cost-effective.

- **Upskilling and reskilling**: Invest in training existing staff on relevant GenAI concepts and tools. This leverages existing talent and reduces reliance on expensive external consultants.

Infrastructure and Pricing Models

- **Cloud-based infrastructure**: Use cloud platforms for training and deploying GenAI models. Cloud services offer scalability, cost optimization features, and pay-as-you-go models, reducing upfront infrastructure investments. However, the pricing models will vary based on the types of services, the models, usage patterns, service levels, and other requirements.

- **Evaluate pricing models**: Choose pricing models based on usage. For high usage, consider time-based or fixed pricing instead of token-based pricing. While token-based pricing might be suitable for proof of concepts, it can rapidly expand with increased usage, especially for summarization of large documents.

- **Resource optimization techniques**: For self-hosting PSOs, implement strategies such as model compression and efficient resource allocation within the cloud environment to minimize unnecessary compute costs without sacrificing performance.

Model and Infrastructure Selection

- **Rightsizing your FM**: Choose the appropriate model size based on your use case. Larger models handle complex tasks better, but you may not always need the biggest one. Different FMs have strengths in different tasks; selecting the optimal size and type is crucial for cost optimization.

- **Choosing the optimal infrastructure**: Explore a broad range of GPUs and accelerators to balance performance and cost. Use dynamic resource provisioning for UI, API gateways, orchestrators, and language models. If using cloud services, leverage auto-scaling features or serverless options for usage-based cost reflection.

Additional Considerations

- **Optimizing prompts**: Prompt engineers should optimize prompts to reduce the number of round trips between the orchestrator and the language model, minimizing processing costs.

- **Reduce development time with better tools**: Use fully managed cloud services to accelerate the development cycle and reduce overhead associated with infrastructure management.

- **Continuous monitoring and optimization**: Continuously monitor the GenAI application's performance to identify further cost optimization opportunities, such as adjusting resource allocation, fine-tuning models, or exploring alternative solutions.

By implementing the preceding strategies, PSOs can optimize the cost of GenAI implementation while maximizing its potential benefits. While GenAI costs might seem significant, it is essential to consider the potential long-term cost savings through increased efficiency, reduced manual labor, and improved service delivery. Therefore, conducting thorough ROI analyses to quantify the expected cost savings and benefits associated with each GenAI project provides a path forward. Using this approach, PSOs can prioritize the right initiatives, enabling improved service delivery and constituent engagement.

9.7 Conclusion

Implementing GenAI successfully in public sector organizations requires careful consideration of several technical and operational aspects. By understanding the anatomy of a GenAI application and the role of each component, organizations can make informed decisions about the tools, technologies, and processes needed to ensure optimal performance, scalability, reliability, security, and cost-effectiveness.

The Foundation Model Operations (FMOps) methodology provides a structured approach to managing the unique challenges associated with Foundation Models, enabling standardization, governance, responsible AI practices, and efficient model deployment and monitoring.

Achieving high performance, scalability, and reliability involves optimizing various components, from front-end applications and orchestration layers to data management strategies, model deployment techniques, and robust infrastructure. Leveraging cloud-based services can significantly simplify this process, but organizations opting for self-hosted open source models must carefully consider factors such as hardware acceleration, memory, storage, and deep learning frameworks.

Security and privacy are paramount considerations, particularly in the public sector. Implementing robust access controls, content filtering, encryption, network segmentation, and continuous monitoring across all components is crucial to mitigating risks and building trust in GenAI solutions.

While the potential benefits of GenAI are significant, the associated costs can be substantial. Careful planning, leveraging open source models, optimizing infrastructure and pricing models, and implementing cost-effective strategies for human expertise can help organizations maximize the value and efficiency of their GenAI investments.

As GenAI continues to evolve and mature, organizations that proactively address these implementation, operational, and cost considerations will be better positioned to harness the power of this transformative technology, driving innovation and improving public services for their constituents.

CHAPTER 10

Summary, Emerging Industry Trends, and Next Steps

In this book, we provided an overview of Generative AI (GenAI) and how it can improve the operations of public sector organizations (PSOs) and enhance the quality of services they provide to constituents. From content generation to conversational interfaces, from content summarization to reporting and analytics, GenAI has the tremendous potential to drive efficiencies and enhance mission delivery across diverse domains.

As we discussed earlier in the book, GenAI is a groundbreaking technology that builds upon AI/ML concepts such as neural networks and transformers. The unique generative abilities unlocked by Foundation Models (FMs), including GPT, Meta, Claude, and Stable Diffusion, represent a significant technological leap. The public sector, with its massive scale, complex operations, and constituent-centric mission, stands to reap the benefits of GenAI.

© Sanjeev Pulapaka, Srinath Godavarthi and Dr. Sherry Ding 2024
S. Pulapaka et al., *Empowering the Public Sector with Generative AI*,
https://doi.org/10.1007/979-8-8688-0473-1_10

10.1 Recap and Summary

Let's recap a few of the key takeaways and considerations from throughout the book:

- **GenAI is an opportunity for PSOs**: FMs such as GPT, Claude, Gemini, and Stable Diffusion offer PSOs the opportunity to enhance constituent services and improve employee productivity.

- **Strategy is key for successful GenAI adoption**: GenAI is a transformative change that necessitates strategic commitment from leadership, workforce reskilling, agile processes, securing adequate funding, and fostering a data- and AI-driven culture. We provided a blueprint for developing GenAI strategy and the key factors for a successful GenAI implementation, including people, processes, technology, and data.

- **Responsible AI**: The nature of GenAI presents some unique challenges when it comes to privacy and security. Organizations in the public sector should adhere to responsible AI practices and include those throughout the implementation.

- **Architecting, building, and delivering GenAI applications**: GenAI brings its nuances for architecting, building, and delivering applications. We discussed these nuances, including key concepts such as prompts, prompt engineering, model parameters, Retrieval-Augmented Generation (RAG), and fine-tuning techniques. We also explored several high-level application architectures that could be used for different types of tasks and methods to improve the accuracy and performance of FMs.

- **Deeper dive into PSO use cases**: GenAI can be used for a broad spectrum of public sector use cases. We reviewed specific examples, including generating documents, reports, code, and multimedia content, powering intelligent chatbots and enterprise search, automating content summarization, and enhancing reporting, business intelligence, and analytics capabilities. We also reviewed domain-specific implementations, including procurement, public communications, research, constituent services, and more.

- **Implementation and operations**: We highlighted the challenges and considerations for deploying GenAI solutions at scale in the public sector. We outlined the best practices, performance, scalability, and best practices for safe, secure, and responsible implementations. We also emphasized the importance of managing operations and costs for successful GenAI adoption.

Looking ahead, the GenAI journey for the public sector is only beginning. As the technology matures, new frontiers will open up, further enhancing human productivity through intelligent task automation. Ultimately, GenAI can provide the opportunity to reshape and transform public sector service delivery while upholding ethical values.

10.2 Emerging GenAI Industry Trends

The GenAI landscape is rapidly evolving, so it's worthwhile to keep abreast of trends that may be beneficial to the public sector, each of which we discuss in this section.

10.2.1 GenAI Models for Audio and Video

In this book, we mainly focused on LLMs, text to image, and MLLMs that handle both text and images. However, the capabilities of FMs are now extending far beyond this. They can now comprehend, synthesize, and transform data across various modalities, including audio and video. These are called "Omni" models.

Imagine a scenario where a Spanish-speaking constituent calls a government agent to apply for healthcare benefits. Here's how a more advanced GenAI system could possibly assist:

- **Understanding needs**: The system analyzes the conversation, grasps the constituent's requirements, and retrieves relevant insurance plan recommendations (text-to-text, Spanish-to-English translation).

- **Form processing**: The constituent uploads eligibility documents. The GenAI system extracts information and uploads it to a database (image to text).

- **Transcript generation**: The system analyzes the audio call and generates a complete transcript (audio to text).

- **Performance feedback**: Based on sentiment analysis of the audio, the system creates a training video (text to video) for the agent. This video highlights areas for improvement, such as active listening, patience, and inclusive language usage.

Real-world examples are already emerging. WhisBERT,[1] for example, is an FM that handles audio prompts. OpenAI's Sora[2] model can create realistic and imaginative videos from text instructions. OpenAI's GPT-4o[3] (o stands for "Omni") accepts as input any combination of text, audio, image, and video and generates any combination of text, audio, and image outputs.

These advancements demonstrate the immense potential of GenAI in public service settings. By seamlessly processing and understanding information across various modalities, GenAI can significantly enhance constituent interactions and streamline government operations.

10.2.2 The Rise of Domain-Specific GenAI Models and Applications

As discussed earlier, techniques such as RAG and fine-tuning allow FMs to adapt to specific domains, including healthcare or finance. This trend of building domain-specific GenAI models and applications is gaining significant traction, as seen with BloombergGPT[4] in finance and Paige AI's GenAI applications[5] for cancer diagnosis.

The focus on domain-specific models is a great sign. It ensures that GenAI applications deliver the level of accuracy required for each industry, ultimately providing constituents and customers with the most relevant and reliable services.

---

[1] https://aclanthology.org/2023.conll-babylm.21.pdf

[2] https://openai.com/index/sora

[3] https://openai.com/index/hello-gpt-4o/

[4] www.bloomberg.com/company/press/bloomberggpt-50-billion-parameter-llm-tuned-finance/

[5] https://paige.ai/

10.2.3 Increased Focus on Small Language Models

While Large Language Models (LLMs) and Multimodal Language Models (MLLMs) have dominated our discussions, Small Language Models (SLMs) offer a distinct advantage in their suitability for specific tasks.

Think of them as smaller, more focused versions of LLMs such as ChatGPT. While not as versatile, SLMs excel in tasks such as text classification and sentiment analysis, making them ideal for specific applications.

SLMs require less training data, computational power, and memory, leading to faster training, deployment, and inference. This allows the models to run on edge devices, expanding their reach. They are easily adaptable to domain-specific tasks, which means PSOs can fine-tune them for their unique needs, ensuring greater accuracy and relevance. Some examples include DistilBERT, Orca 2, and GPT Neo.

As computing power and training options evolve, SLMs are poised to become even more powerful and versatile, offering a valuable alternative (to LLMs) for public sector GenAI applications.

10.2.4 LLM-Powered Autonomous and Multi-agent Ecosystem

We explored the concept of "agents" and their potential for complex reasoning and multistep problem-solving. This trend takes a significant leap with the emergence of multi-agent ecosystems. These involve multiple GenAI agents collaborating to achieve a common goal, each specializing in specific roles within the application.

Imagine a multi-agent system designed to personalize the learning experience for students in public schools. This system could use various agents to achieve this goal:

- **Content generation agent**: This agent leverages an FM to analyze student data (e.g., past performance, learning style preferences) and generate personalized learning materials. This could include customized study guides, practice questions, or even adaptive learning content that adjusts to the student's progress.

- **Tutoring agent**: Another agent, powered by another FM, could act as a virtual tutor. It can answer student questions in real time, provide explanations in different formats (text, audio, video), and offer personalized feedback based on the student's understanding.

- **Assessment agent**: This agent uses an FM to analyze student performance data (such as test scores and assignment grades) and identifies areas where the student needs additional support. It can then trigger the content generation agent to create targeted learning materials or connect the student with the tutoring agent for further assistance.

- **Progress-tracking agent**: This agent uses an FM to monitor the student's overall progress, identifies learning patterns, and adapts the system's recommendations accordingly. It can also generate reports for teachers and parents, providing insights into the student's learning journey.

This multi-agent ecosystem demonstrates how GenAI can be used to personalize education in the public sector. By collaborating, these agents can create a dynamic and adaptive learning environment that caters to individual student needs, ultimately improving educational outcomes.

10.2.5 Embedded GenAI Capabilities to Improve Productivity

One of the emerging trends is the evolution of products and tools in the industry that have inherent GenAI capabilities embedded within them. These products improve productivity by helping with day-to-day tasks, such as email, document generation, document analysis, reporting, business intelligence, routine question and answer for human resources operations, and policies, among others. Prime examples include Microsoft Office Copilot[6] and Amazon Q.[7]

10.2.6 The Rise of Intelligent Devices: GenAI at the Edge

The advancements in AI and Internet of Things (IoT) have significantly changed the livelihood of constituents, from smart cities to remote patient monitoring in healthcare. Edge computing allows tasks to be implemented at the edge devices, such as surveillance cameras, sensors to monitor critical infrastructure (such as water processing plants or power grids), and field operation devices used for investigations, agriculture inspections, and so on.

With the concept of GenAI at the edge, ideally, the inference is implemented on the edge devices improving the efficiency to take timely action as opposed to bringing the data back to the cloud or a data center for inference. GenAI at the edge is applicable to many use cases within the public sector, including

---

[6] https://blogs.microsoft.com/blog/2023/03/16/ introducing-microsoft-365-copilot-your-copilot-for-work/
[7] https://aws.amazon.com/q/

- **Smart cities**: Traffic management, public safety, and surveillance; edge devices can analyze videos and images to alert law enforcement in real time.

- **Emergency and disaster management**: GenAI can be used in edge devices such as weather sensors to enable quick responses in disaster situations.

- **Environmental monitoring**: Edge devices equipped with GenAI can help detect anomalies in water quality and air pollution and generate alerts in real time.

- **Education**: GenAI can be used at edge devices such as interactive white boards to further enhance the student experience. As an example, GenAI can generate content on the white board based on simple annotation or a prompt. Additionally, GenAI-powered smart boards can interact with students or teachers to provide question and answer capabilities with chatbot capabilities.

Overall, the convergence of GenAI and IoT is an important emerging trend with great potential for PSOs to improve the mission outcomes with faster decision-making, low latency, and improved overall efficiency.

10.3 Next Steps

This book has provided a comprehensive overview of GenAI's potential to transform public sector operations. Here are some key steps to further your understanding.

10.3.1 Expand Your Knowledge

Gain a deeper understanding of the entire GenAI ecosystem, from applications to infrastructure. Identify key service providers, available models, tools, and technologies for successful implementation. Develop a tailored GenAI strategy based on the framework provided in Chapter 3. This blueprint will guide your organization's successful adoption of GenAI.

10.3.2 Prepare Your Workforce

If you are a leader or manager, prioritize workforce readiness. Explore available training and certification paths for your team. Consider free online courses offered by Microsoft, Google, and Amazon as starting points.

10.3.3 Identify and Prioritize Use Cases

Inventory potential GenAI use cases that align with your organization's mission objectives. Begin planning pilot projects to experiment and evaluate the technology's impact.

10.3.4 Experiment and Learn

There are a number of mechanisms for you to experiment and learn – some of these are as follows:

- **Sandbox environment**: Set up a controlled GenAI environment to experiment with one or two primary use cases, initially focusing on internal applications. This sandbox will help you understand how various technologies and components interact within the GenAI ecosystem.

- **Model selection**: Choose the most suitable model from available options based on your specific needs.

- **Prompt engineering**: Optimize your results by mastering the art of prompt engineering.

- **Fine-tuning**: Tailor models for your specific domain (such as finance, tax-related information) to enhance accuracy and relevance.

- **RAG mastery**: Explore advanced Retrieval-Augmented Generation (RAG) techniques for handling complex document structures including images, tables, and fine print.

- **Production readiness**: If satisfied with the pilot project's outcomes, plan for production implementation of a selected set of use cases. The concepts discussed in Chapter 9 are provided as guidelines to help ensure production-grade GenAI applications are created.

10.3.5 Stay Ahead of the Curve

We cannot emphasize enough how important it is to stay up to date in this rapidly changing field. Keep reading and reviewing articles and discussions in the news about the field. For example, a new technique called Retrieval-Augmented Fine-Tuning[8] (RAFT) has been showing some promise. Try to attend industry conferences and subscribe to journals and

---

[8] https://arxiv.org/abs/2403.10131

newsletters. Deepen your understanding by actively following the latest research developments in GenAI through platforms such as NeurIPS[9] and arXiv.[10]

The guidance and topics covered in this book give you the platform necessary to embark on a successful journey with GenAI. We wish you the best in unlocking this technology to ultimately improve public service delivery and constituent engagement.

---

[9] https://neurips.cc/
[10] https://arxiv.org/

List of Foundation Models

A.1 GitHub Repositories

- **uncbiag/Awesome-Foundation-Models**: This curated list categorizes FMs for various tasks like vision and language.

- **facebookresearch/foundation-models**: Meta AI maintains a list of their open source FMs on GitHub.

A.2 Articles and Websites

- **Deci AI – Top 10 List of Large Language Models in Open-Source**: This article provides an in-depth look at several prominent open source LLM FMs.

- **Medium – List of Open Sourced Fine-Tuned Large Language Models (LLM)**: This Medium post compiles a list of open source fine-tuned FMs.

© Sanjeev Pulapaka, Srinath Godavarthi and Dr. Sherry Ding 2024
S. Pulapaka et al., *Empowering the Public Sector with Generative AI*,
https://doi.org/10.1007/979-8-8688-0473-1

A.3 Search Engines

- **Hugging Face Model Hub**: Filter by "Open Source" under the "License" option to find a wide range of open source FMs.

- **Papers With Code**: Use keywords like "foundation model" and "open source" to discover research papers and repositories related to open source FMs.

A.4 Foundation Model Provider Reviews

- www.gartner.com/reviews/market/generative-ai-model-providers

APPENDIX B

List of Tools for GenAI

B.1 Open Source Vector Databases

- **Qdrant**: Offers a user-friendly interface and API for production-ready similarity search and vector storage

- **Milvus**: Scalable and flexible with strong community support, popular for large-scale deployments

- **Weaviate**: Focuses on knowledge graphs and semantic search and integrates well with GraphQL ecosystems

- **Faiss**: Primarily a library for similarity search, often embedded in larger systems

- **ClickHouse**: Offers vector functionalities alongside its traditional time-series capabilities

- **OpenSearch**: Extends its full-text search engine to include vector similarity search

- **Vald**: Focuses on high-performance similarity search with efficient nearest neighbor retrieval

- **ScaNN**: A C++ library known for its fast and accurate approximate nearest neighbor search

- **Pgvector**: Integrates vector functionalities into the PostgreSQL database management system

© Sanjeev Pulapaka, Srinath Godavarthi and Dr. Sherry Ding 2024
S. Pulapaka et al., *Empowering the Public Sector with Generative AI*,
https://doi.org/10.1007/979-8-8688-0473-1

B.2 Commercial Vector Databases

- **Pinecone**: Google's cloud-based vector database, excels in scalability and performance
- **Chroma**: Offers advanced search functionalities like multimodal and subgraph search
- **Deep Lake**: Focuses on large-scale scientific and research applications
- **Elasticsearch**: Adds vector search capabilities to its popular search engine platform
- **Vespa**: A platform from Exosphere focusing on real-time search and recommendation systems
- **Marqo AI**: Offers a cloud-based AI platform with strong search and ranking capabilities

B.3 Built-In Vector Search Capabilities

- Amazon OpenSearch
- Amazon DynamoDB
- Amazon RDS for PostgreSQL

B.4 Orchestrators

- LangChain
- LlamaIndex

APPENDIX C

List of GenAI Applications

C.1 Code Generation and Completion

- **GitHub Copilot**: The pioneer in this space, Copilot suggests code completions, functions, and even boilerplate code based on your current context, significantly boosting your coding speed and efficiency.

- **Tabnine**: Another powerful assistant, Tabnine predicts your next coding moves with context-aware suggestions, automatic fixes, and instant access to relevant documentation and examples.

- **Kite**: Get direct in-editor support with Kite's AI assistant. It provides explanations, definitions, and even alternative solutions for your code snippets, making learning and exploration seamless.

- **Code With Me**: This unique platform pairs you with an AI copilot that can not only complete your code but also explain its decisions and reasoning, fostering a collaborative learning environment.

© Sanjeev Pulapaka, Srinath Godavarthi and Dr. Sherry Ding 2024
S. Pulapaka et al., *Empowering the Public Sector with Generative AI*,
https://doi.org/10.1007/979-8-8688-0473-1

- **Amazon CodeWhisperer**: Amazon CodeWhisperer is an AI-powered productivity tool for the IDE and command line that generates code suggestions based on comments and existing code.

C.2 Design and Optimization

- **DeepCode**: Apply the power of AI to code reviews with DeepCode. It scans your code for security vulnerabilities, code quality issues, and potential bugs, ensuring robust and well-maintained codebases.

- **Sourcegraph**: Navigate and explore your codebase effortlessly with Sourcegraph's AI-powered search and visualization tools. Discover connections between functions, track code evolution, and identify potential refactoring opportunities.

- **Replit**: Leverage AI for experimentation and exploration with Replit's cloud-based coding platform. Build, test, and deploy various code structures and configurations instantly, fueled by AI-powered suggestions and debugging tools.

C.3 Personalized Learning and Assistance

- **ChatGPT for Coding**: Fine-tuned for programmers, this Large Language Model can answer your coding questions, explain complex concepts, and even generate basic code snippets, acting as a virtual coding tutor.

- **Codurance**: Get real-time coding guidance and feedback from experienced developers via Codurance's innovative platform. Combine human expertise with AI suggestions for a powerful learning and problem-solving experience.

- **CodeSignal**: Hone your programming skills with CodeSignal's AI-powered coding challenges and interview simulations. Test your knowledge in various languages and frameworks, receiving personalized feedback and insights for improvement.

APPENDIX D

Model Cards

Many model cards use the following format:

Model Details Name of the Model

Model Developers Owner of the Model

Variations parameter sizes, information on pre-trained and fine-tuned variations.

Input Type of input: image or text

Model Architecture type of model architecture - transformer architecture, etc.
||Training Data|Params|Context Length|GQA|Tokens|LR

Model Dates Dates when the model was trained.

Status Offline / online

Licenses type of license

Research Paper details of paper on which the model is based if any

Where to send questions or comments about the model Instructions on how to provide feedback or comments on the model

© Sanjeev Pulapaka, Srinath Godavarthi and Dr. Sherry Ding 2024
S. Pulapaka et al., *Empowering the Public Sector with Generative AI*,
https://doi.org/10.1007/979-8-8688-0473-1

#**Intended Use**
Intended Use Cases Use Case tasks & language

Out-of-scope What the model is not intended for

**Note: Additional Notes

Hardware and Software
|Time (GPU hours) |Power Consumption (W)|Carbon
Emitted(tCO$_2$eq)|
|---|---|---|---|

Training Data
Information on source of data and when the training was cut off

Evaluation Results
Results for the models on standard academic benchmarks.

#Ethical Considerations and Limitations**
Associated Risks and considerations for use

See the following model cards for examples:

- https://github.com/facebookresearch/llama/blob/
  main/MODEL_CARD.md

- https://github.com/openai/gpt-3/blob/master/
  model-card.md

Others may provide similar information in different formats, such as
the following examples:

- www-cdn.anthropic.com/files/4zrzovbb/website/
  bd2a28d2535bfb0494cc8e2a3bf135d2e7523226.pdf

- https://aws.amazon.com/machine-learning/
  responsible-machine-learning/titan-text/

APPENDIX E

COTS Tools for Program Management Tasks

- Copilot for Microsoft 365: `https://adoption.microsoft.com/en-us/copilot/`

- Cohere Generate: `https://cohere.com/generate`

- Synthesia: `www.synthesia.io/`

- Claude: `www.claude.ai`

- GPT for Google Sheets and Docs: `https://workspace.google.com/marketplace/app/gpt_for_sheets_and_docs/677318054654`

© Sanjeev Pulapaka, Srinath Godavarthi and Dr. Sherry Ding 2024
S. Pulapaka et al., *Empowering the Public Sector with Generative AI*,
https://doi.org/10.1007/979-8-8688-0473-1

Open Source Deployment Frameworks

- www.langchain.com/langsmith
- www.bentoml.com/
- https://github.com/ray-project
- https://github.com/bentoml/OpenLLM
- https://modal.com/
- https://github.com/jina-ai/jina
- https://ollama.com/

© Sanjeev Pulapaka, Srinath Godavarthi and Dr. Sherry Ding 2024
S. Pulapaka et al., *Empowering the Public Sector with Generative AI*,
https://doi.org/10.1007/979-8-8688-0473-1

APPENDIX G

Code Generation Tool Evaluation Framework

Category	Subcategory	Evaluation Aspect	Description
I. Needs Assessment	NA	Identify tasks and workflows	Analyze current processes to identify repetitive tasks for automation.
		Technical requirements	Consider programming languages, frameworks, and integration needs with existing systems.
		Security and compliance	Evaluate security measures, data privacy practices, and compliance with regulations.
		Ethical considerations	Assess potential biases in generated code and ensure alignment with ethical guidelines.

(continued)

© Sanjeev Pulapaka, Srinath Godavarthi and Dr. Sherry Ding 2024
S. Pulapaka et al., *Empowering the Public Sector with Generative AI*,
https://doi.org/10.1007/979-8-8688-0473-1

Category	Subcategory	Evaluation Aspect	Description
II. Tool Evaluation Criteria	A. Functionality		
		Supported languages and frameworks	Does the tool support the organization's programming languages and frameworks?
		Task coverage	Does the tool address the identified tasks and workflows effectively?
		Code quality	Does the generated code meet coding standards and maintainability?
		Accuracy and reliability	How accurate and consistent is the generated code?
		Customization options	Can the tool adapt to specific coding styles and project requirements?
	B. Usability and integration		
		Learning curve	How easy is it to learn and use the tool effectively?
		Integration with existing workflows	Does the tool integrate with existing development tools and IDEs?

(continued)

Category	Subcategory	Evaluation Aspect	Description
		Documentation and support	Does the vendor provide adequate documentation, tutorials, and support resources?
	C. Security and compliance		
		Data security practices	How does the tool handle sensitive data and ensure its security?
		Compliance with regulations	Does the tool comply with data privacy regulations and security standards?
		Auditability and traceability	Can the tool track code generation history and provide audit logs?
	D. Cost and licensing		
		Pricing model	Is the pricing structure clear and transparent?
		Licensing options	Does the licensing model meet deployment needs (on-premises, cloud based)?
		Scalability	Can the tool scale to accommodate future growth?

(continued)

Category	Subcategory	Evaluation Aspect	Description
III. Evaluation Process	Shortlist potential tools		Identify tools that meet the organization's core requirements.
		Trial and proof of concept	Conduct trials with shortlisted tools on specific tasks.
		Benchmarking and performance testing	Compare the performance and accuracy of generated code.
		Security assessment	Evaluate the security posture of the tool and its compliance.
IV. Selection and Implementation	NA	Select the most suitable tool	Select the tool that best aligns with the organization's needs.
		Develop implementation plan	Define training, deployment strategy, and integration plan.
		Monitor and evaluate	Continuously monitor performance, user feedback, and impact on development processes.

Index

A

Adoption (GenAI)
  data-driven organization
    challenges, 79
    concepts, 76
    decentralized data
      architecture, 79
    governance, 80, 81
    initiatives, 77
    phrase, 77
    privacy/security, 78
    quality, 78
    sharing/integration, 79
    volumes/types, 78
  framework implementation
    data-driven
      organization, 76–81
    people perspective, 60–64
    perspectives, 58, 59
    process, 64–69
    technology evaluation/
      selection, 69–76
  implementation
    application/orchestration
      layer, 52
    business problem
      definition/planning
      stage, 50, 51
    continuous monitor/audit/
      fine-tuning, 53
    data collection/
      processing, 51
    evaluation/selection, 51
    fine-tune/training, 51
    lifecycle, 49, 50
    production deployment, 53
    test, validate, monitor, and
      audit, 52
  people perspective
    ACOEs, 61
    framework outlines, 60
    gradual culture
      evolution, 61
    innovative technologies, 62
    objective key results
      (OKRs), 61
    policies, 60
    roles/responsibilities, 62–64
    strategic vision, 60
    workforce enablement/
      education/
      transformation, 64
  process perspective, 64
    automation, 69
    bias and fairness
      detection, 67

S, T

U

V

W, X, Y, Z